Andrew Parker was born in England in 1967. He received his Ph.D. from Macquarie University in Sydney while working in marine biology for the Australian Museum. He became a Royal Society University Research Fellow at Oxford's Department of Zoology in 1999, and is an E. P. Abraham Research Fellow of Green College, Oxford, a Research Leader at the National History Museum, London and a Research Associate of the Australian Museum and University of Sydney. He has published numerous scientific papers on topics as diverse as optics in nature, biomimetics and evolution, and is the author of the acclaimed *In the Blink of an Eye: How Vision Kick-Started the Big Bang of Evolution*. He lives in Oxfordshire.

Also by Andrew Parker

In the Blink of an Eye

seven deadly colours

The genius of nature's palette and how it eluded Darwin

Andrew Parker

fP

FREE PRESS

First published in Great Britain by The Free Press, 2005
An imprint of Simon & Schuster UK Ltd
A CBS COMPANY
This edition published, 2006

1 3 5 7 9 10 8 6 4 2

Simon & Schuster UK Ltd
Africa House
64–78 Kingsway
London WC2B 6AH

www.simonsays.co.uk

Simon & Schuster Australia
Sydney

A CIP catalogue for this book is available
from the British Library.

ISBN: 0-7432-5941-6
EAN: 9780743259415

Typeset by M Rules
Printed and bound in Great Britain by
Cox & Wyman Ltd, Reading, Berks

To Natalia

Contents

List of Illustrations

Integrated illustrations

Plate section

1. The pygmy sea horse camouflaged against coral. © Scott Tuason/imagequestmarine.com

2. ULTRAVIOLET: An ultraviolet photograph of a primate's head. No colours visible to humans are used in this photograph.

3. VIOLET: South East Asian butterflies. The upper butterfly is a female Malayan Eggfly; the lower butterfly is a male Blue Crow.

4. BLUE: 'Sea sparkles' (dinoflagellates) in a cup of sea water. This photograph was taken by Peter Herring in complete darkness and without a camera flash. © Peter Herring/imagequestmarine.com.

5. GREEN: An enigmatic blue frog, *Rana caerulea*, and a green tree frog of Australia.

6. YELLOW: A crest feather of the Sulphur-crested cockatoo, taken in darkness using an ordinary camera flash (left) and an ultraviolet-only flash (right). The ultraviolet flash is not visible to humans, yet we see *both* feathers appearing yellow.

7. ORANGE: A milk snake, with a twist in its colour tale.

8. RED: A dragonfish of the deep sea, cheating the 'rules' of colour. © Peter Herring/imagequestmarine.com.

The lowest and vilest alleys in London do not present a more dreadful record of sin than does the smiling and beautiful countryside.

SIR ARTHUR CONAN DOYLE, *The Adventures of Sherlock Holmes*
– The Copper Beeches (1892)

Preface

This is the second book of a trilogy on the evolution of vision and how colour and other visual adaptations found today are relevant to the Big Bang of evolution (the 'Cambrian explosion'), some 540 million years ago. The first book, *In the Blink of an Eye*, laid down the idea that the very first eye on Earth triggered an arms race that continues today. *Seven Deadly Colours* aims to provide a flavour of the *sophistication* of colour in animals – one part of that arms race where an animal's visual appearance on Earth and its eyes are key weapons.

The eye was Charles Darwin's nemesis. He worried about the road of eye evolution – how the eye could possibly evolve as a series of small steps, especially when some potential steps fail to add little or nothing to sight (at least relative to the energy needed to make the additional structural and chemical modifications). I will deal with this problem, however, in my next book (the last of the trilogy). The solution, it so happens, is relevant also to the Cambrian explosion, so completing the circle. *Seven Deadly Colours* will deal only with Darwin's 'extreme perfection' issue with the eye. To solve *this* problem, I derive clues from the colour of animals today.

In Darwin's mind, the eye was the height of natural sophistication, a masterpiece of nature. By assigning perfection Darwin simultaneously cast doubts over its maker – was the eye just too accomplished to have been conceived by evolution? he thought. Quite a problem for a man toying with the idea of natural selection in the face of a religious onslaught. But Darwin was less complimentary of nature's hues, considering colour as the poor relation in the visual family but also completely independent of eyes. Taking this view, evolution simply 'does its best' to adapt animal coats to the presence of retinas. An 'anything-is-better-than-nothing' approach as far as colour is concerned. *Seven Deadly Colours* will get to the bottom of colour in animals, and consider the subject as part of a visual 'whole'.

The machinery behind the colour of animals today is diverse. But if these colour mechanisms are as sophisticated as the eye itself, then there was no reason for Darwin to worry after all. Although the eye is an obvious weapon, could colour be an equally effective counter-weapon and even 'put one over' on the visual system? Is colour actually as *deadly* as the eye? If so, *the eye could not be perfect after all*. Then all would become well for the Darwinian theory of evolution.

Surprisingly, considering that we view many of the machines for colour every day, they are not well known. Simply, we are blind to what is happening in the skins, furs, feathers, shells and spines of animals. Maybe plain old paints have been responsible for the poor public relations afforded nature's colours. The array of colours on an artist's palette is all down to one mechanism – that of pigments. Sounds boring, I know. But there is more to pigments than one may think, and I will devote one chapter to them. Although only one. On the *painter's* palette lie violet pigments, blue pigments, green pigments, yellow pigments, orange pigments . . . On *nature's* palette are violet nano-optics, blue bioluminescence, green pixel permutations, yellow fluorescence, orange optical illusions . . . there are even colours we cannot see! If only Darwin had been aware of the virtuosity and sophistication of *colour* in nature, and balanced this against the respect he held for the eye, his complete confidence in evolution might have been restored.

This book is made up of both acute technical (mathematical and chemical) details and more general accounts of natural history, including scientific exploration, animal behaviour, ecology and evolution. The level of attention demanded of the reader by these two domains contrasts highly. During the writing of this book it became clear that one domain had to give somewhat, since each could occupy the entire book on its own. I chose to abridge the technical details, which I have reduced to their minimum without compromising on scientific accuracy. Nevertheless, those details are sufficient to explain how the different colour factories found in animals cause colour. They can, however, be skipped if the reader prefers, since the stories of natural history are unaffected by such interludes (although the descriptions of what light is and how the visual system works, which feature early on in this book, are worth digesting to augment the later chapters). I hope that this system, whichever way it is used, helps to provide fluid accounts of colour in each species selected. And there lies the second problem faced while writing this book – species selection. There are so many astonishingly coloured animal species in nature, each with their own fascinating natural histories, yet I had space only for seven. At least I could cover the basic factories for producing colour within them all. It should be borne in mind, however, that the cases of colour and its remarkable adaptation provided in this book are by no means the whole story. They are merely representatives of the myriad battles between the sophistication of colour and eyes in the rainforests, coral reefs, sandy deserts, deep seas and mountain-tops of the Earth.

My research on colour in nature began at the Australian Museum, Sydney, in 1993 and continues today. In particular I work on structural colours, as described in the Violet chapter of this book. I have many people to thank for their help in this subject, from physicists and chemists to biologists and artists. At the beginning I remember the fascinating literature that swayed me from a biodiversity study of seed-shrimps towards the optical devices that emerged on one of their antennae, the literature of Sir Eric Denton, Michael Land and Peter Herring, who had helped classify structural colours in nature and explain such cases as the silver colour of fish, the iridescence of shells and the changing colours of cuttlefish. These biologists helped me shape my study within the seeming infinity of nature's colour. My research progressed through collaboration with optical physicists, particularly Ross McPhedran, David McKenzie and Maryanne Large (Sydney University), Zoltan Hegedus (CSIRO, Sydney) and Mike Gale

(CSEM, Switzerland). I have enjoyed working with these people and many other biologists, physicists and engineers since then (particularly Chris Lawrence at QinetiQ), and the subject of optical devices in nature has provided a continuous source of amazement. On the theme of research I am indebted to the BBSRC and ARC in Australia and to the BBSRC, NERC, EPSRC and QinetiQ in the UK for funding my work, although above all to the Royal Society (London) for my current Fellowship. Additional support was provided through my college positions in Oxford – an Ernest Cook Fellowship at Somerville College and an E.P. Abraham Fellowship at Green College – where my eyes are opened to subjects other than biology.

The acknowledgements provided in *In the Blink of an Eye* apply equally to this book, particularly the thanks given to Jim Lowry, Penny Berents and Pat Hutchings (Australian Museum), Noel Tait (Macquarie University), Marian Dawkins and Paul Harvey (Oxford University). Thanks also go to the researchers whose work I describe in this book, including André Nemésio, Walter Boles and Mark Norman, for their patience during my torrent of questions. Additionally, I thank Michael Bishop for organising an eight-lecture course at the Sydney Grammar School in 2003, where I first presented 'Seven Deadly Colours'. But special thanks for help and support in compiling this volume are due to my family, and to my editor, Andrew Gordon, for his continual advice and always intelligent and valuable questioning in cross-examining my scientific and literary ideas.

Introduction

I sell here, Sir, what all the world desires to have – POWER.

MATTHEW BOULTON (British engineer), 1776

Dispelling some myths

Vision is perhaps the overall dominant sense in an animal world shaped by senses. Eyes confer a monumental advantage on their host species, and could be considered the object of all animals' evolutionary desire. But at the same time eyes exact much in return in terms of the energy required to make them and the sizeable brain capacity and nervous wiring that must be borne by the animal's body. Because these 'costs' often cannot be met, eyes that form images have evolved in only six of the thirty-eight phyla or categories of multicelled animals. Yet the monumental advantage they bestow is reflected in the success of those six phyla – over 95 per cent of species belong to them. Today over 95 per cent of all animals on Earth have eyes that form images. Related to this, unlike the other universal senses of hearing, smell and taste, the stimulus for vision – light – is ever present. The combination of these two facts means that almost wherever we go, somewhere our image will lie on a retina. And not necessarily a human retina. Like all animals, we are under constant surveillance. But should we care? Well, certainly, especially if that retina belongs to a lion while we're on safari on the Serengeti, or a brown bear while we're trekking through the Rockies.

And that sentiment can be extended to most other animals, where predation is a bigger issue.

Most predators use vision to find their prey, and most prey endeavour to visually detect their predators. But evolution has dealt both predators and prey visual weapons other than eyes in their arms race. Some have been handed a tactic to avoid visual detection – camouflage. Camouflaged body shapes, stealthy behaviours and, most universally, inconspicuous colours. Others have found themselves with bodies that noticeably spell danger or aggression, laying down the gauntlet to their predators. Sharp spines protrude from bodies, and swords slash through the air around them. But then colour is equally employed to signal a readiness for battle. Colour – perhaps the dominant weapon in the fight for animal survival. And that's just natural selection.

Colour is perhaps even better known for its role in sexual selection. The evolution of proficient mate attraction, resulting in more offspring, works in parallel with natural selection within nature's arms race. As a group, birds perhaps employ colour most spectacularly of all, largely because they can escape most predators through flight. For many birds, the advantages of visual conspicuousness in attracting a mate outweigh those for discreetness. But the balance is a delicate one. Whichever way you look at it, predators and vision go hand in hand – colour is a deadly business.

We are all aware of the common if not legendary stories of colour in nature. There's the dull green of a leaf insect that achieves precise camouflage, the stripes of a tiger that can break up its outline and extinguish the tiger shape, the evolution of the peppered moth from grey to black as its background trees became polluted with soot, and the colour change of a chameleon to match any environment it wanders into. At the other end of the scale, dull green grasshoppers reveal shocking pink wings to startle their predators during escape, hummingbirds flash metallic-like reds to attract a mate from some distance, and angler-fishes dangle glowing blue lures to attract inquisitive prey. But occasionally an exceptional story is uncovered, where one must become immersed deeply in the science and history of the specific subject to benefit from the twists and turns, shocks and surprises, and wonder of the solutions to the problems. This book tells seven such stories. They cannot be summarised in a sentence; not even a para-graph.

We may also feel that we know the basics behind the cause of colour in nature. The reason for this can be summed up in a single word – pig-ments. Often when we think of colour, we think of art – a human response to vision. Just ask an artist how each colour on a canvas is formed, and oil paints packed with chemical pigments will provide an answer for the entire spectrum. But evolution, in its response to vision, has been far more varied and ingenious. Behind the scenes of a colour lies a microscopic factory, responsible for the light that leaves an animal's body or artist's paint. Yes, pigments represent one colour fac-tory. Pigments are molecules, and their factory involves electrons that whiz around when fed on some colours. But they are unable to con-sume other colours, which are rejected into the surroundings and maybe into eyes. The pigments' factory is the only one on an artist's palette, but it is just one of seven factories in animals. Admittedly, it is the commonest colour factory in nature, luring us into thinking we know it all already. But just wait to find out what else is involved. Each chapter in this book, while telling its own exceptional story of colour, will expose the remaining factories. And how could one debate colour without serious consideration of vision? So to balance colour, there's another theme running through this book; a gravely important theme that involves the eye.

Darwin's dilemma

On 1 July 1858 Charles Darwin and Alfred Russel Wallace made the first public statement of their theory of evolution by natural selection. This theory had occurred independently to each of them. Darwin first recorded his argument in a *Sketch* in 1842, followed by a more finished *Essay* in 1844. The *Essay* is a shorter, simpler and more direct version of the celebrated *On the Origin of Species* of 1859, but has been granted equal importance by evolutionary biologists. Wallace reached similar conclusions fourteen years after Darwin, but before Darwin had published his own results. Darwin and Wallace presented their theory in a joint publication, in the *Proceedings of the Linnean Society*, in 1858. However, it is Darwin's *Essay* that first records his battle with an item apparently inconsistent with his theory. It must have been troubling him for some time, and may have been one reason why he did not publish the *Origin* sooner.

Darwin considered evolution a gradual process and had no problem with the idea that insect legs could eventually evolve to be jaws. Here, numerous gradual changes from leg to jaw can be easily predicted and, considering both the vast (geological) time periods and space on Earth available for this to happen, conceived as practical in the human mind. This is reassuring. The list for similar cases is long – think of an elephant's tusk with its origin in an ordinary tooth, or, moving in the other direction, the strong, aerodynamic wings of a seabird that became the more vestigial, hydrodynamic wings of a penguin. In his study in Down House, just south of London, Darwin observed the many beetles collected during his five-year expedition on board the *Beagle*. The study is north-facing, which in England is synonymous with dullness, or at least the lack of direct sunlight. The gloomy furniture – deep brown wood and dark green cloths – compounded the murkiness, but the available light was enough for Darwin to scrutinise his beetles.

Encased in glass-fronted oval boxes, or shallow cabinet drawers, each beetle was suspended in mid-air by a single pin driven through its abdomen, so preserving its brittle appendages. Darwin would remove the beetles by holding the ends of the pins, and transfer them to a 'microscope', in the broadest sense of the word. His desk was equipped with a magnifying glass at the end of a telescopic arm, but also with the very latest microscope – block of wood as stand, vertical brass rod as support, and two glass lenses held in sequence horizontally. This was barely an improvement over Robert Hooke's invention of the seventeenth century, but at least one lens was movable – the microscope could focus.

Darwin noted the slight variations in beetle leg shapes and sizes as he grouped the most similar species together. The metallic sheen of particular specimens appears to have escaped his attention, or at least the commitment of his (valuable) time. Maybe if he had taken the same beetles into his garden on a summer's day, he would have been compelled to describe and think more carefully about the dazzling colours passing through a spectrum as they departed the exoskeletons and infiltrated the atmosphere. An effect akin to a thick layer of dust blown from a jewel would have struck Darwin on his journey from darkness to sunlight. Darwin, alas, remained in the dark and went on to play devil's advocate to his own logic.

There is one body part that Darwin singled out as different; a problem of the highest order for his theory of evolution. This part lies beyond the limits of the theory, as it is acceptable to humans. While Darwin's thoughts were flowing smoothly from case to case of easily conceivable, gradual transition, and hurtling towards an unquestionable theory of evolution, he would at some stage return in need of an explanation to that body part – the eye. Precisely, in the *Origin* Darwin wrote:

ORGANS OF EXTREME PERFECTION AND COMPLICATION.

To suppose that the eye, with all its inimitable contrivances for adjusting the focus to different distances, for admitting different amounts of light, and for the correction of spherical and chromatic aberration, could have been formed by natural selection, seems, I freely confess, absurd in the highest possible degree.

An important note: please always bear this extract in mind throughout this book.

Clearly, Darwin felt the need to forestall the potential criticism that would accompany the publication of his theory because he could not apply it to the eye. By taking this precaution, he made the most of his time on a stage that would soon become filled by the more vocal resistance that lay just around the corner. But it is the title of the above extract that is most interesting – 'Organs of Extreme Perfection and Complication'. The term 'extreme perfection', where it relates to the eye, will be addressed throughout this book.

It is imperative to understand that an eye is an organ that forms a visual image on a retina, as opposed to a light sensor (also known as a light-sensitive patch or light receptor), which simply detects the direction of light. This distinction should not be forgotten, particularly where the evolution of vision is concerned.

It is not surprising that Darwin singled out the eye. Yes, it appears a most sophisticated and complicated organ, and, while paying only compliments, why not go all the way and call it perfect? Well, there is evidence in another of Darwin's works to suggest that this may not be the case.

In his *Essay* of 1844, Darwin listed another interesting example that could be taken as contradictory to his theory. He noted that:

> The nature or condition of certain structures has been thought by some naturalists to be of no use to the possessor, but to have been formed wholly for the good of other species . . . [for example] certain fish to be bright coloured to aid certain birds of prey in catching them.

He went on, later in the same paragraph, to defend his theory with:

> No doubt one being takes advantage of qualities in another, and may even cause its extermination; but this is far from proving that this quality was produced for such an end . . . the bright colours of a fish may result from exposure to certain conditions in favourable haunts for food, *notwithstanding* it becomes subject to be caught more easily by certain birds.

Here, Darwin allows a glimpse into his mind, where animal colours were not a tool he had chosen to construct his evolutionary theory. He made broad generalisations based on some unfortunate cases – he assumed that colour is incidental in (some) conspicuous fishes. Darwin cannot be blamed for this, of course. His task was formidable, with so little biological foundation to build on. And to make things worse, that foundation included some very weird and wonderful theories on colour in nature. There was the idea that animals could somehow photograph their environment on to their bodies and so exhibit the same colours. Then there was the theory of colour evolution, where animal colour must always pass through the spectrum in the direction from red to violet on the road to an adaptive hue. In other words, blue was more costly than red because it resulted from a lengthier transition.

Our blood is red, and we all know that this *colour* is without a function or purpose. Clearly Darwin was thinking along these lines when he explained that fish were simply unfortunate to be so brightly coloured because this attracts the attention of their bird predators. Here evolution is not considered perfect; rather the positive attributes of species outweigh the negatives. Fine, but Darwin went too far with his explanation for the hues of a fish. He suggested that the chemicals or light conditions in the water had altered the fish's colours, making it maladapted. Only a detailed study of the colour factories in the fish's skin, along with its precise appearance in the eyes of its predators, can reveal the truth. As this book will demonstrate, that fish was colour-adapted after all.

It is equally unfortunate that Darwin regarded the eye as the pinnacle of evolution – *nothing* gets past the eye, he thought. He considered that the fish's skin had lost the visual battle with the predator's eye. A century and a half of scientific experimentation later, we can provide alternative views that are certainly closer to the truth. Hence the question to be answered in this book, while investigating nature's colour factories, is 'Was Darwin right to consider the eye perfect?' But returning to the factories, the remainder of this chapter will provide a feel for what is meant by a colour factory, and for how each type can possibly be different when they all, apparently, have a similar result – the manufacture of colour.

Monet's flaw

A colour comes with a *range* of qualities, and not all colours seen in nature have the same qualities. That would seem to fit with my previous announcement that not all colours result from the same type of factory. This is not so simple as different chemicals causing different colours. By different factories I mean microscopic machines that are poles apart. As a means of demonstration of this contrast, we could consider Claude Monet's shortcomings when he began painting pheasants.

The hard winter of 1879–80 made painting out of doors difficult for Monet, so he turned, for a brief period, to making still lifes, which he had not entertained since 1872. At this time he resided in a comfortably-sized, box-shaped house at the end of a row of four, in Vétheuil, on the Seine, just outside Paris. The weather became bitterly cold – temperatures fell to a record –25°C – but after snowfalls the skies cleared, and as the wooden shutters were swung open the sun streamed through the large six-panelled glass windows.

Monet began his indoor series with a trio of fruit paintings. Lining up melon slices, apples and grapes on his table, he proceeded to paint them with faithful reproduction of their hues. He was able to do this because the pigments behind the colour of fruit are comparable to those in artists' paints. So no problem for Monet – he simply mixed red and blue pigments on his palette and applied the purple blend to the canvas. Just like a real grape, his rendition looked the same from all corners of the room. Indeed many objects we encounter in our daily lives, including most animals and plants, are coloured as a result of pigments.

Monet replaced the fruit on his table with freshly killed game birds, laid on their backs. He must have complained as the sun, reaching a critical point in the sky, suddenly poured through his windows and flooded his room with light. Closing the louvred shutters served only to shred the sunbeam, casting unworkable stripes on his subjects. This lighting problem was quite out of character for Monet, since he had built a reputation upon recording various sunlight conditions so realistically. That was even the secret of his realism – although individual objects in his work are often fuzzy, one can instantly identify with his landscapes as if standing within them because of the spontaneity of the light effect. Sometimes the same landscape was depicted at different times of the day, where the sun's colours served as a clock. Variation in light was Monet's forte.

Preferring the even illumination, Monet threw the shutters open and allowed the sun into his studio. But it took just a quick glance at the game birds on the table for a new problem to emerge – a problem never encountered in his landscapes. The plovers with their pigmented plumage were fine, but the pheasants . . . the pheasants displayed shining green heads and golden bodies. As Monet moved around the pheasants, parts of their bodies would vary from invisible to positively dazzling. Colours began to turn on and off – the birds looked quite different depending on where he stood in the room. The intense golden reflections caused all else in his field of view to fade into nothingness, and mesmerised Monet like a rabbit caught in headlights. Then a step to the left and . . . the pheasants continued to blaze but were now unmistakably green. Monet moved his head around the paints on his palette, and alas, their colours were static. How could he reproduce this dynamic effect on canvas armed only with pigments that display the *same* effect when viewed from any direction? Could he unleash a hitherto hidden character of his oils?

Well, he couldn't. Monet eventually compromised by colouring just part of each pheasant's head with a green streak while the remainder of the head plumage faded to black. This at least indicated that something was peculiar. But the green streak on the canvas was really nothing special, fooling us only momentarily by its contrast with the black that surrounded it. As one walks past the painting, the green streak stays where it is. But as Monet moved around the pheasants, the position of the green part of the birds moved with him, step by step, feather by feather.

This is one of those colours in nature that has different qualities from pigments. The on-off metallic green flash of a pheasant's head cannot be found on an artist's palette. It requires an altogether different factory from the pigment machine. And this green flash is only one deviation from the effect of the assumed universal pigments. In total there are six deviations from pigments in nature, resulting from six more factory types, and they will all be covered in this book. Conveniently, there are as many types of colour factories in animals as there are colours in the rainbow. And that's a gift I have grasped with both hands – each chapter in this book will cover a different colour *and* a different colour factory. This skeleton will be fleshed out with stories to settle Darwin's nerves over the so-called 'perfect eye', and so light, colour, vision and evolution will intertwine. But just before linking these subjects, we need some basic knowledge of one of them – light. What, exactly, is light? A good place to begin searching for an answer is the home of Isaac Newton.

Newton's groundwork

Two centuries before pheasants troubled Monet, Isaac Newton sat in a much darker room on the other side of the English Channel. Between 1665 and 1666, with the plague in its prime, Newton escaped from his study at a precarious Cambridge to the isolated safety of his home at Woolsthorpe in Lincolnshire. At the age of only twenty-two, and in addition to laying the scientific foundations of maths and astronomy, he began experimenting with prisms. Not one prism, as others had played with before, but two. And it was the second prism that revealed the true secret of sunlight, or 'white'.

It was already known that sunlight could be split into the colours of a rainbow by passing it through a prism. But earlier investigators believed that the prism itself *altered* the sunlight in some way as it passed through the glass, so the character of sunlight was changed. Newton arranged his sparse, dark room with a table in the middle. On the table he aligned, from right to left, a magnifying glass (a lens) and a prism. To the left of the table a white board was set up, reaching almost to the ceiling, with a series of small holes lining up vertically. To the left of this lay the second prism, mounted directly behind the lowest hole in the board. Nothing else lay between the second prism and the white wall of the room behind it. Newton waited.

The sun came round the corner of the house and eventually streamed in through the window at the right-hand side of the room. The sunlight was visible as a beam from Newton's view, side-on to his apparatus, as it illuminated the dust in the air. In order of events, the beam collided with the lens at a shallow angle and became redirected and focused towards the first prism. It then passed through the prism where it divided up into a spectrum, and struck the large board over a range of angles – red beams lit the board lower down, violet beams higher up, with a complete rainbow in between. Red light struck the board at precisely the position of the lowest hole, and so passed through it. On reaching the second prism, this red beam was further bent at precisely the same angle the first prism had bent it. But, against all understanding of the time, after transiting the second prism the red beam became . . . a red beam. Remarkable! The second prism had not altered the red beam. So prisms *do not* alter the nature of light!

Newton rethought the mechanics of a prism. White light from the sun became a series of colours, but the colours could not be divided further. Sunlight, therefore, is actually a mixture of all the colours in the spectrum, Newton deduced. And of course, he was right. Simultaneously he had also promoted the spectrum to a new level of importance – it was a general property of white light and not an artefact of a prism. Now this continuum of merging colours, sprawling from violet to deep red, required some sort of classification.

Different accounts exist of why Newton gave the rainbow, or white light spectrum, seven colours – violet, indigo, blue, green, yellow, orange and red. One account involves his interest in musical harmonies, where there are seven distinct notes in the scale. Newton, the story goes, proceeded to divide up the spectrum into spectral bands with 'widths' (ranges of wavelengths for each colour) corresponding to the ratios of the small whole numbers in the scale. Another account involves the culture of the time, in which the number seven had magical or biblical significance. Either way, Newton's seven colours are not the best choice. If we are to divide up the spectrum into the colours we perceive, although strictly the colours do merge to form an infinite sequence, then today we prefer to omit indigo from Newton's categorisation. Indigo is not really seen as a separate colour. This leaves the modern spectrum with the order: violet, blue, green, yellow, orange, red. Six colours. So why is this book entitled *Seven Deadly Colours*? This is because the spectrum is actually broader than Newton realised. Without giving further explanation at this stage, there is an additional colour known as 'ultraviolet'.

So Newton explained that sunlight contains seven colours, although he misunderstood what light was. He maintained that light was a stream of corpuscles. The first objection to this idea was that two people would not be able to look into each other's eyes because the corpuscles passing to and fro would hit each other and fall to the ground. The great Dutch physicist Christian Huygens was first to discover (the predecessor of) the wave nature of light in 1678. It was later assumed that waves of light existed in air or another medium that acted like an elastic solid. The even greater Scottish physicist James Clerk Maxwell nullified inconsistencies in this idea in 1880 with his theory that light is an 'electromagnetic' wave.

Colour – the product of light and vision

The sun continually emits waves, of different cycle lengths, of electro-magnetic radiation. I will just take a moment, or four short paragraphs, to expand upon this sentence because it is so imperative to this book. 'Waves' in this case refers to the displacement of a medium, such as the molecules in air, in a wave-like pattern. Stretch out a rope and flick one end with the wrist – a wave will travel along the rope until it meets an object in its way. It oscillates up and down. The term 'wavelength' refers to the distance for the wave to complete one cycle – the distance from one high point on the rope to the next, as the wave flows along it. It is helpful to imagine the pattern formed by the waves travelling across the rope as if frozen in time. That's how we consider light waves – as 'profiles'. The wave on the rope may have a wavelength of about a metre; the wavelengths of light waves are measured in nanometres (they are about 2 million times smaller).

The term 'electromagnetic' refers to the composition of the wave – it is essentially a pair of waves. It has an electrical component and a magnetic component, like two ropes stretched out together where each is flicked by the wrist in a different direction (one up and down, the other left to right). They are identical in their profile and move at the same speed, but at 90° to each other. While an electromagnetic wave travels across this page, the electric component oscillates from top to bottom while its magnetic counterpart flows into and out of the page.

Electromagnetic waves contain energy, which originates at the sun. The sun emits waves of a continuous range of energies, and this translates to a range of waves of different wavelengths. The greater the energy, the shorter the wavelength – to make many short waves along a length of rope, the wrist must flick harder (using more energy) than it would to make fewer, longer waves. Within the broad range of electromagnetic wavelengths emitted by the sun lies a relatively narrow section that is prominent at the Earth's surface – this section is known as 'light'. And within the light region, violet is associated with the shortest wavelength and most energy, and red with the longest wavelength and least energy.

Finally, we consider a single wave as a ray, and a bundle of rays as a beam. The rays within a beam may have different wavelengths and are not usually orientated all in the same plane. As a beam travels across this page the electrical component of some of its rays may oscillate from top to bottom of the page, with others slightly out of and into the page. Indeed, the rays of some beams may continually change their planes of oscillation so that they spiral across the page. And that's the basics of light, which is worthwhile digesting.

These paragraphs on the nature of light fall within the realms of physics and demand an element of concentration compared to, say, the accounts of Darwin's experiments or Monet's compositions. The narratives of the following chapters in this book unite history, biology and art and could be described as friendlier, although at times they are more involved. The central stories flow in their own right, through evidence, red herrings and twists, from problem to solution. But the

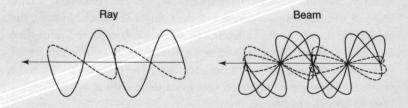

Ray   Beam

Figure 1.1   The electromagnetic nature of light – a ray, and a beam with three rays (of the same wavelength). The solid line represents the electric component, the dashed line the magnetic component, which displace air molecules at 90 degrees to each other.

technical details of colour factories add spice to the mix. These details are included in the chapters, albeit stripped to their bare bones and confined to their own, separate sections, all beginning with the subheading 'Details of the colour factory'. So the choice is offered to skip the simple physics or chemistry behind the colour factories and move directly to the subheading 'Back to the story'. Returning to the portrayal of colour in this chapter, I should move on from an understanding of light, and give a few more technical details.

Newton did recognise that 'the rays are not coloured'. Colour, in fact, is not a property of light at all but is purely a property of the visual system – the eye and the brain, with all its wiring between. Without a visual system, there would be no such thing as colour.

Our vision begins when a ray of light strikes our cornea – the outer surface of the eye. Next the ray meets the lens of the eye where its path is bent, and finally ends its existence on the retina, at the back of the eyeball. Here, the ray may be absorbed by a cone cell, a more sophisticated form of those simple light receptors found within light-sensitive patches in some eyeless animals. Rod cells also exist in the retina, but these are responsible for detecting low levels of light. The cones specialise in bright light and 'colours'.

Cone cells contain visual pigments made up of two parts – an opsin (protein) bound to a chromophore (a vitamin A derivative). When struck by a package of light of a specific wavelength, the opsin and the chromophore interact, form an excited state, and then return to normal again in the time taken (in femtoseconds, i.e. 10^{-15} seconds) for an atom in the chromophore to complete a single vibration. Simply, the colourless opsin can be thought of as becoming coloured, and then bleached. During bleaching, a chemical is activated that causes a signal in the nerve fibres that join to the cone cells. The end result within the retina is that a nerve impulse is sent towards the brain, as the opsin is reset to absorb another photon.

Each cone cell is serviced by its own nerve cell. A nerve cell consists of a rounded cell body, containing the nucleus, and long, thin cellular extensions known as nerve fibres. Nerve fibres can reach up to 1 metre in length, and can connect to each other to form a chain or network of nerve fibres. This network, however, is not permanent, and juxtaposed

fibres can compete for connections depending on their activity requirements. There is a degree of plasticity in the system.

The part of the brain used for vision is called the visual cortex, and
in humans this lies in a part of the brain at the back of the head. The
last nerve fibre in a chain delivers the signal collected from the cone cell
here. The visual cortex is where images are constructed, often in colour.
Here is the first stop along the visual pathway where colour can be said
to exist – colour is born in the visual cortex. Objects around us are not
coloured, they just reflect different combinations of wavelengths. *The
brain gives each wavelength its own colour.* Throughout this book I
will sometimes cut corners by referring to an object as coloured – you
will know what I really mean.

We have three different visual pigments in the cone cells of our retinas, which form the basis of our colour vision. One visual pigment
absorbs wavelengths between 400 and 500 nanometres, although it
absorbs best at 425 nanometres, which the human brain translates to
'blue'. Cone cells with this pigment are called blue cones.

The second visual pigment in humans absorbs wavelengths between
455 and 605 nanometres, although maximally at 530 nanometres,
which becomes translated to 'green'. Cone cells with this pigment are
called green cones.

The third visual pigment in humans absorbs wavelengths between
485 and 700 nanometres, although maximally at 560 nanometres,
which becomes translated to 'yellow' (the wavelengths approaching
700 nanometres are translated to 'red'). Cone cells with this pigment
are called red cones.

Using these three visual pigments, any wavelength between 400 and
700 nanometres can be seen by humans. This is the range of the human
visual spectrum, as illustrated by Newton, only now we have explained
it.

The conclusion to be drawn from all of this is that we live in a virtual reality world, where the images we perceive actually do not exist as
such. Objects, including animals, surround us. They consist of matter
of some description; matter that has shape, mass, texture . . . maybe
also movement. But one character they do not possess is colour.

The film *The Matrix* had a point. This rightly considered the basis of

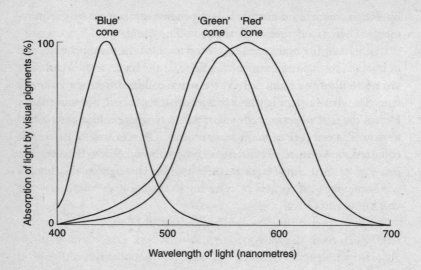

Figure 1.2 The absorption of light by the three visual pigments in the human retina.

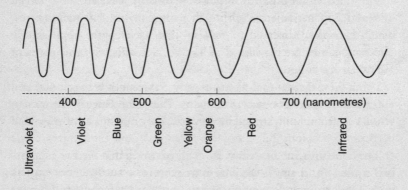

Figure 1.3 The region of the electromagnetic spectrum known as 'light' and its translation to colours by humans.

all human senses as detectors that feed electrical signals to the brain, which the brain converts to a recognised odour, texture, flavour, temperature, musical note or colour. In *The Matrix* the world had been destroyed by war and became an unpleasant place to live. So humans were wired up to computers. The nerves leading from the brain to a detector were intercepted and connected instead to a computer. The human detectors became redundant as the computer fed its own stimuli to the brain. The humans were unaware of anything feigned, and occupied a congenial land as designed by the computer (the Matrix). But even under natural conditions, for animals with colour vision, life *is* surreal.

Energy pinball

Eyes belong to living animals and, as you would expect, show signs of life themselves. I have noted the reactions that spark changes in the chemicals of the retina, and electrical impulses that whiz along nerve fibres. But colour-causing devices are not static either. These devices, embedded within the skins, scales, feathers and shells of animals, involve activity on a 'nano'-scale. Electrons can jump around within single atoms, or they can leap from atom to atom within a molecule, all at supersonic speeds. Sometimes chemicals react to form not only new compounds that affect light but also light itself. Then there are the microstructures that act like pinball machines, causing light waves to bounce around inside them. Alternatively waves and patterns of waves can play tricks on our eyes and fool us into thinking there is something ahead that is really not there. In all cases the end result is that the device sends a collection of wavelengths in the direction of our eyes, which we perceive as a colour. But I will say no more about these microscopic colour factories in this chapter, other than that an artist's paints encompass just a single mechanism. Each of the subsequent chapters will cover a different colour of the rainbow. They will include different groups of animals, each engaged in a different part of nature's global arms race, where the eye is a key weapon. But they will collectively embrace the full spectrum of production lines that relentlessly churn out the different coloured effects we perceive around us, every day, from dawn to dusk, and even during the night. The skins and shells of animals, their living tissues, hard external parts, secretions, will all be scrutinised atom by atom.

Selection pressures are those invisible forces in the surroundings of a species that drive and shape evolution. The range of colour factories in animals is a consequence of the selection pressures established by the eye. Eyes exist on Earth, so individuals continuously leave images of themselves on retinas. This has been the case for over 500 million years. Any genetic mutation that causes the eye of a predator to be fooled, with the result of a longer life expectancy, will be maintained within the species. Mutations that cause changes in shape, behaviour and also colour are included here. Leaf insects and peacocks would never have evolved if eyes did not exist. And if the eye is in fact not perfect, evolution will probe away to find its weakness in the form of colour evolution in the animals it views. Everything is moving in this arms race, from the electrons and light waves that bound with energy to cause colour, to the genes that code for those electrons and pinball machines of light. The paints, monotonous in their lack of diversity, which solely occupy an artist's canvas, are supposed to depict animal shells, skins and feathers. But what they actually represent are the bustling, dynamic activities, albeit at a microscopic scale, relentlessly restrained with the bodies of life forms, fighting for the cause of their hosts' survival. If Darwin had known this he would not have made his remark about the unfortunate, uncomplicated colours that just pop up effortlessly in conspicuous fish and are received gratefully by the elegant eyes of birds of prey.

'NanoCam' – a fictional nano-camera

For the purposes of this book, technology will take a step further towards the invisible, crossing the line between fact and fiction, and conceive 'NanoCam' – the world's first camera capable of filming structures and events at the sub-micron or nano scale – including the scale of the wavelength of light. The ingenious parts of NanoCam are the two nano-sized fibre-'optics' that can even handle X-rays; electromagnetic radiation that is much smaller than light waves.

The optics of NanoCam are so thin that they are not visible to the naked eye, and are flexible and lie juxtaposed. They consist of the radiation source transporter, the 'source optic', and the radiation collector, the 'camera optic', which can be steered. An injector is included in the set-up to feed the delicate optics in among the atoms of the specimen under examination, be it solid, liquid or gas.

The camera optic plugs in to a stack of equipment that includes a 'monitor wall'. There are nine individual monitors in the monitor wall, each tuned to display pictures produced from different wavelength ranges, with visible light in the top left monitor and X-rays in the bottom right. Never-before-seen events such as light waves crashing into molecules, molecules interacting during chemical reactions and, right at its lower limits, electrons moving within their orbitals around the nuclei of atoms, can now be filmed by the camera optic as they are illuminated by the source optic. And all events are displayed to the human eye, as they happen, on the monitor wall.

Unbelievable? Yes, of course, but if it were real what a wonderful tool to investigate the cause of colour in nature or even art. And since that is what this book will do, NanoCam will be brought to virtual life and employed.

NanoCam can also examine paintings, to really make comparisons of artists' colours with those of animals. But I will say no more about either at this stage, other than that NanoCam will reveal that nature's colour palette is far more sophisticated than most of us believe, and the eye less so. These facts, respectively, are the cause of Monet's and Darwin's troubles. To reveal them is the aim of this book. NanoCam, therefore, will become a most useful tool. Now to put it to work in solving seven cases of colour found in nature that do not appear to follow the rules of an artist's paints. Should be fun.

COLOUR 1

ultraviolet

The problem:
Why do kestrels hover over motorway verges, where their prey is well camouflaged?

The first colour in nature's spectrum can be difficult to get to grips with. It will eventually provide the solution to the above problem, but it can never really be fully conceived. Why? Well, the world was not designed by man or with humans in mind, so colour in nature is not primarily for our benefit. We must ask, 'How do other animals view the world?' There is our world for us, and their world for them. The central problem in this chapter will immerse us headfirst into 'theirs'.

The puzzle, as it would appear to us, is that kestrels hovering over motorway verges are a familiar sight, but in this location their prey is perfectly camouflaged against the background undergrowth. So how can kestrels possibly feed in their chosen grounds when, apparently, they can't see their food? After all, vision is the main prey-finding sense for kestrels.

Like costly advertisements for beer or chocolate that appear on TV, the proof that the actions of kestrels are effective is time. Beer and chocolate commercials continue to influence our decision when faced with a bar or a shop shelf today just as they did decades ago. In other words, they work. The cost of advertising campaigns is more than recovered by profits from additional sales. If they did not work, they would not exist today. Kestrels, in my experience, are as common

Figure 2.1 A kestrel resting and hovering.

beside motorways today as they were twenty-five years ago. So whatever it is kestrels are doing, that must be working too. And we do
know what they are doing. The question is *how* are they doing it?
How do kestrels spot the voles they prey upon when the voles appear
indistinguishable from their background?

For birds of prey, kestrels operate rather close to the ground.
Buzzards and vultures soar high on the thermals that facilitate gliding.
Such energy conservation affords a modest energy input – they can
feed when food presents itself. Kestrels are in the market for smaller
prey, even in relation to their body size. Due to size alone, the voles
they consume could not possibly be seen from great heights, and as a
result kestrels are forced closer to the ground. Hovering is their alternative to soaring. But hovering *is* a drain on energy. Wing muscles are
great energy-consumers, so kestrels must be efficient vole-consumers.
And observe a kestrel hovering for just a couple of minutes and you

are likely to witness a kill. Before long its wings will fold back, the kestrel will nose-dive, wings will be deployed like a parachute at the last second and the predator will return to the air with a vole in its talons.

The reason for hovering is to keep a still head. When the eye and object viewed are nearby, movement becomes a particular problem for vision, whether it is eye or object that is moving. The faster the movement, the bigger the problem. Imagine gazing through a train window. We unconsciously turn our heads to remain (successfully) focused on distant fields, but we have no solution to the blurring of nearby hedgerows. Flies manage to keep their heads still as their bodies rotate. This behaviour can be observed most dramatically in the stalk-eyed flies – flies with eyes at the end of very long stalks protruding sideways from their heads. These flies can rotate their bodies gradually, but during a complete 360° turn their heads make just a few jerky movements to catch up. This way, their eyes are held still for most of the movement, allowing the flies to see during rotation. Similarly, walking pigeons rhythmically flick their heads forward while their bodies travel smoothly, overtaking the head at times. Again, the head and eyes are motionless for much of the journey. So by remaining still in the air, kestrels can avoid the problems inherent in movement and fix their vision on a single point. But what point? The point where a vole happens to be resting? If so, our problem remains. How does the kestrel know where the vole is resting?

Mammals missed out on evolving bright colours. Somewhere on the evolutionary tree, around the point where mammals branched off from the reptiles, some 230 million years ago (before dinosaurs), colour vision was lost. The first mammals evolved to see in black and white.

Early mammals were not top predators, as are some of their better-known progeny today. Quite the contrary, they were small shrew-like forms (hairy lizards may be a better description) that engaged in the only way of life available – they lay low. Mammals lay low for most of their history, until the demise of the dinosaurs 65 million years ago. Early mammals, certainly, were impending food for the large amphibians and reptiles of their day, and accordingly their evolutionary balance tipped towards inconspicuousness.

I refer to the conspicuous–inconspicuous balance. There are advantages in both possibilities for appearance . . . but there are drawbacks too. A bright, colourful coat will attract individuals of the same species from afar, which in turn may lead to increased breeding and population size. Success by 'duplication', (population) safety in numbers, and all that. On the down side, what is attractive to friends will tend to be a beacon to foes. At its worst, a great, glowing beacon. Now take the inconspicuous option, and the inverse problem. The flip side of camouflage to a predator is concealment from potential mates. How can one find a mate when they are all well camouflaged? Avoidance of predators can translate to a struggle for reproduction. Hence the evolutionary balance.

Evolution works under the influence of selection pressures – those invisible forces in the physical environment (such as temperature) and biological environment (such as competitors) that act ultimately on genes. In the case of colour, the eyes of enemies and the eyes of one's own kind convey strong selection pressures. Ways of attracting mates or avoiding predators become the responses open to a species (a species, in most cases, is a group of like individuals that will reproduce in their natural environment). The response with the *most* positive outcome for the survival of the species will be that employed (over time the genetic mutations that afford this outcome will be retained within a species, while others will be lost).

The first mammals could compete with their neighbouring *small* amphibians and reptiles for the food available. No problem with regard to intake. The decisive attribute for species survival, then, was a birth rate that exceeded the death rate. The first mammals achieved this primarily through targeting the death rate. They evolved low-profile body shapes, and drab colours that blended in with their background to provide visual camouflage against predators. The large amphibian and reptile predators of these mammals most probably saw in colour. Maybe the predators' skins displayed brightly coloured patches, for warning or mating purposes. Early mammals, on the other hand, were condemned to the environmental gutters – the dark, dingy undergrowth of 'forests'. Colours here were mainly browns and greys. Accordingly, early mammals were likely brown or grey.

Early mammals fed mainly on insects. Although some insects were brightly coloured, in dimly lit mammal-land their colours would have faded to grey (almost) – the reason for which will become evident as this chapter unfolds. Even worse for colour, the mammals' chief advantage over their amphibian and reptile competitors for insects lay in their blood, their *warm* blood. This allowed them to stay active at night, and as a result early mammals were largely nocturnal. With potential mates and food all appearing dull, the selection pressures for colour vision in mammals were not strong, and they evolved retinas with rods – rod cells that provide black and white vision, rather than cone cells that distinguish 'colours'. On land, black and white vision beats colour vision hands down at night.

By surrendering the colour-sensitivity of their eyes, early mammals improved their resolution. Rather than evolving retinas packed with cone cells that would each detect light rays of specific wavelengths, they became filled with cone cells where each would sense any wavelength. One rod performed the job of three cones, so reducing the 'grain size' of images captured, albeit in black and white. Black and white camera film carries this advantage over colour film for the same reason.

There were two positive outcomes for this selection. First, smaller patterns and shapes could be resolved with black and white vision. This made mammal vision more sensitive to movement, useful for spotting insects. Second, they could see better in low light conditions – three times better, using the comparison here. They could be active at dawn, at dusk and even under moonlight. Advantages aside, without colour vision selection pressures for bright coats were minimal. The point of this reasoning is that early mammals were an unremarkable brown, like the general colour of their background. Extrapolating this logic, voles have continued the trend today.

The origin of colour in the course of seeing

Voles are *brown*. What does that mean, exactly? In the previous chapter light was described as electromagnetic waves, and the colour of light was dependent upon the cycle length of those waves. A spectrum of colours can be seen in a rainbow – the colour in natural sunlight. Voles are lit by natural sunlight, but brown is not a colour in the spectrum. So what is it?

A spectrometer is a machine that measures wavelengths of light. The light to be tested enters the spectrometer, through a small hole or probe, and strikes a diffraction grating. A diffraction grating, in this case, is a small, flat, rectangular plate with thousands of parallel, microscopic grooves etched into its surface. When white light meets the grooves, it is reflected, but not equally. Blue waves are reflected at a different angle to green waves; yellow waves at a different angle to red waves. The overall effect is that white light is split up into its spectrum. A hologram as found on a credit card contains diffraction gratings. Note how different colours are reflected into different directions, even though white light illuminates the card from only one direction. If, however, only blue light is used to light up a diffraction grating, only blue light will be reflected, and there will be nothing seen in the directions otherwise occupied by green, yellow and red. Now it becomes clear how a spectrometer works. Place sensors in the directions where blue, green, yellow and red wavelengths emerge from a diffraction grating, and each of those components of white light can be measured separately. Measured for intensity (the number of light rays every second), that is, compared to each other.

Let's see what happens when a spectrometer is aimed at a vole in sunlight. A sector of the light rays leaving the vole's fur passes into the opening of the machine, just as they may pass into our eyes to be seen. The light rays leaving the vole's fur strike the diffraction grating.

As they reflect from the grating, the rays splay out into a spectrum, like that leaving a prism. They keep travelling within the spectrometer until they meet a sensor – the blue wavelengths meet the blue sensor and so on. Here they end their existence, as the sensors take a measure of their individual intensities. A chart is produced revealing the proportion of each wavelength present in the light leaving the vole's fur. There is a peak in this chart.

It so happens that all wavelengths in white light were reflected from the vole's fur, violet through to red. But the violet, blue, green, yellow and orange wavelengths were reflected much less than the reds. The peak in the reflectance chart is in the red. Why, then, do we not see the vole as red?

First, to establish if spectrometer readings can really help to explain our vision, compare a spectrometer to an eye. The spectrometer does not have a brain – its whole is equal to the sum of its parts. The visual complex, on the other hand, has a whole that is greater than the sum of its parts, thanks to an additional processing component – 'intelligence'. The comparison of a spectrometer to an eye is, consequently, unfair. A better comparison would be a spectrometer to a package of rod cells – one red, one green and one blue.

The region at the back of our eyeball that images are focused on to by our lens is called the retina. It is where the rods and cones are found. The retina alone has the job of data-collection. It *is* comparable to the array of sensors in a spectrometer. Light falls upon the retina, and whatever signal leaves the retina is that which eventually becomes an image in the brain.

We are trichromats. The term 'trichromats' refers to the three types of cone cells we possess in our retinas – one for detecting blue wavelengths, one for greens and one for reds (we also have rods, incidentally). In fact each cone type detects a range of wavelengths, so by having these three cones we can detect any colour in the spectrum. Rather, we can detect light of any wavelength in the spectrum (wavelengths between 400 and 700 nanometres). At this point, we can leave the theory and examine the practice. We have NanoCam at our disposal, and we can feed this into an eye and through the visual hardware to see for ourselves exactly what happens during the process of vision, while viewing a vole.

The microscopic NanoCam is now in place in a human retina. It is lying at the junction of a blue, green and red cone. Its images produced on the viewing monitor show that each cone cell contains a concertinaed outer membrane, with large molecules restrained at certain points. The large molecules appear different in each of the blue, green and red cones – they each have clearly different arrangements of atoms. But what difference does that make? The beam of light heading the way of NanoCam will probably reveal all.

A strong, white light source has just illuminated the vole. The light reflected from the vole's head is hurtling towards NanoCam in the retina. The individual rays within this reflected beam can be seen oscillating as air molecules are pushed aside. Now the rays are close, and strike the cornea, the outer barrier of the eye. Some rays have been lost – reflected away from the eye by its smooth outer surface (smooth surfaces, such as those of a glass window, reflect a significant portion of light – hence you see your reflection in a shop window). But most rays have made it through the corneal boundary and are now within the lens.

Their path has deviated. The lens, and to a lesser extent the cornea, has focused the beam on to NanoCam. The reason for this is the change in 'refractive index'. Albert Einstein, in his theory of relativity of 1905, taught us that light rays travel at different speeds in different media, such as air and water, travelling fastest in a vacuum. So a change in medium, as light passes from air to water, will reduce the speed of the rays, and the effect is that the rays' path deviates sharply at the media interface. A pencil half submerged in water will appear to bend at the water's surface – that is the deviation effect of light rays that leave the submerged part of the pencil and cross the water–glass–air boundaries. The refractive index is simply a measure of how much the rays deviate compared to their passage through a vacuum. Air is assigned a refractive index of one.

Rays leaving the vole's tail strike the lens on its opposite side and also become deviated, focusing on to the retina near NanoCam – within a millimetre. In fact the 'tail rays' cross over the 'head rays' on their journey from lens to retina. The lens, as a result, has focused an image of the entire vole, albeit upside down. The vole is several centimetres in length, but has been converted to less than a millimetre in length on the retina – the small eye can image large objects. Over a lifetime, our eyes will capture around 24 billion images.

All hell breaks loose as the light rays from the vole's head collide with all three cone cells surrounding NanoCam. After traversing a layer of nerves that connect to the cones, the light rays encounter the large molecules in the cone walls and end their life, after a 93-million-mile journey from the sun. This encounter is rather volatile.

The energy in the light rays is relayed to the large molecules, causing the atoms within to shuffle around. The molecules, however, almost instantaneously revert to their original atomic organisation. But the atomic shuffling has a consequence – an electrical pulse flows into the nerve cells in the retina. The large molecules in the blue cone cells absorb only the energy of blue wavelengths – green and red wavelengths cause no reaction in its atoms. The same applies to the large molecules in the green and red cone cells, which fire electrical impulses when contacted only by green and red wavelengths respectively. The first stage of colour vision is beginning to unfold.

Now let's move NanoCam in an attempt to follow the electrical impulses, in this case from a red cone. The physical movement within the large molecules of this cone caused a change in permeability in the entire cone cells' outer membrane, which in turn triggered an electrical impulse. The impulse can be seen on the NanoCam monitor spreading along the cone cell. It is now at the top of the cell, squeezing into the narrow, fibrous terminals. What happens here is important to the workings of any multi-celled animal. The impulse passes through the membrane of the cone cell, leaving it behind, but enters only a 'synapse' (Greek for 'joint'). The beginnings of another cell lie just beyond the cone terminals, but it is not quite touching. There is a small, physical gap – a potential dead-end. Fortunately, though, there are chemical intermediaries called transmitters to carry the impulse across the gap and into the beginnings of the neighbouring cell – the impulse lives on.

The neighbouring cell is rather short and squat at its middle, but again has fibrous beginnings and endings. The impulse can just about be seen to travel through this cell, across another synapse at its end, and into another cell – a retinal ganglion cell. Let's advance NanoCam into the retinal ganglion cell, because this cell appears different.

After bypassing the nucleus within the bulbous beginning of the cell, the impulse takes a 90° turn into the start of a very thin, fibrous region (an 'axon'). NanoCam can detect no end to this fibre. It appears to twist and turn, but not to end. Through its transparent walls, other similar fibres are noted, juxtaposed to this one, forming a bundle of long fibres – the optic nerve. Now we have left the eyeball and are within the visual wiring, or stage two of the three-phase visual system. In the fibre containing NanoCam, the impulse races ahead until soon it is out of sight of the sauntering camera optic. No problem, because from behind comes another impulse, following a second firing of the cone cell after another encounter with a red light ray.

Impulses come and go, overtaking NanoCam as it glides through the nerve fibre. The impulses come and go extremely regularly, over 100 every second. They follow the nerve fibre along its entire length, conveyed to a port in the pink 'blancmange' that eventually surrounds the whole optic nerve. The blancmange is the human brain, our information-processing centre.

The wire-like nerve fibre docks at the sensory relay station (the 'lateral geniculate nucleus') in the geographical centre of the brain. It has dispatched the electrical message generated in the eye all the way to the brain. Now it is the turn of the brain to make some use of that signal.

Moving NanoCam into the brain, the train of electrical impulses it has followed becomes focused once more. The signal remains alive. It continues to hurdle synapses between further nerve cells, possibly with some degree of alteration – that is difficult to observe. Eventually, alas, the signal meets its end.

The electrical impulse makes one last synaptic leap into the bulbous cell body of a nerve cell in the 'visual cortex' of the cerebrum, the area of the brain at the back of the head, directly behind the eyes. *This is where the signal is interpreted as 'red' light.* The conclusion drawn is that the part of humans' environment occupied by the vole's head is reflecting electromagnetic radiation of wavelength 650 nanometres – *red* light. The brain is like a dictionary, translating wavelengths to colours.

To summarise, electromagnetic radiation of a range of wavelengths (sunlight) illuminated the vole, and from one spot (a hair, for instance) on the vole's head electromagnetic radiation of wavelength 650 nanometres emerged. This radiation struck one cone cell in the human observer's retina. The cone cell fired an electrical impulse along a nerve fibre plugged into the brain. Finally, the signal was translated to 'red'. Throughout that whole pathway, colour was not born until it reached the brain. Colour does not exist in the environment, nor does it exist in the eyeball. *Colour exists only in the mind.* Now that abstract statement does make sense.

Repeating this experiment with NanoCam focused on the neighbouring green cone cell in the retina, a similar result is obtained. Similar, but slightly different. Again the cone cell fires electrical impulses as the green light rays leaving the vole's head strike the large molecules in the cells' outer membrane, causing them to momentarily alter their arrangement of atoms and make the membrane permeable. Again impulses follow a path leading to long nerve fibres. But on manoeuvring NanoCam into those green fibres, a difference emerges.

The impulses racing through the green nerve fibre on their way to the brain are just as strong as those in the red fibre. They are also just as fast, causing NanoCam to quickly lose sight of them. But they are less frequent. There were maybe ten impulses in the red fibres to every one in the green fibres.

A third experiment, with NanoCam focused on the neighbouring blue cone cell in the retina, revealed exactly the same result as for the green cone. The blue and green cones are firing impulses at the same rate.

So how does the brain interpret this information? It determines that from the vole's head there are fewer electromagnetic waves of wavelengths 520 nanometres and 460 nanometres than there are of 650 nanometres. Ten times less, in fact. As 650 nanometre waves were interpreted as red, 520 nanometre and 460 nanometre waves are interpreted as green and blue respectively. Similarly, green and blue are born in the brain. But the brain's work is not done yet.

Because these red, green and blue cone cells were so close in the retina, and consequently received light from almost the same part of the vole's body, their signals are combined within the brain to form a single pixel. The red, green and blue colours are combined.

Combine red, green and blue lights and the result is a white light. The vole, however, is clearly not white. Combine lights in the proportion ten parts red, one part green and one part blue, and the resultant colour is . . . brown! This is the brain's result too. The signals for red, green and blue arrived at the brain in these proportions, and the brain combined them to form a 'brown' dot. So brown is not an alternative colour to red, green and blue, in that it has its own visual pathway, it is rather one of the outcomes of a red, green and blue signal (an 'unsaturated' colour). By the same logic, to see a pure (deep and intense, or 'saturated') blue object, the green and red cone cells must be completely inactive. The brain is a rather large dictionary of colour.

There is more to the saga of visual processing in the brain, particularly concerning image formation. The truth is we do not know the end to that story. We believe that the colours assembled in the visual cortex move on to another part of the brain via more nerve connections. This will emerge as an enigma only in the Orange chapter – until then we have all the knowledge needed to solve the specific problems posed in this book. For now, at least, we can move on.

The vole, meanwhile, remains ignorant of the visual activity in the human subject. At the moment, the human is part of an arms race in which the vole is also very much involved. The vole has lost a battle, but not necessarily the war; at least the visual war against humans. This story is far from over.

The vole's technical coloured coat

Half of the whole colour process, the part from where light rays first meet the eye up to the formation of an image in the mind, has now been explained. The other half precedes this, and involves the question: 'How does the vole convert sunlight to brown light?'

A more hard-lined translation of this question is: 'How does the vole convert a beam containing equal amounts of rays with wavelengths from 400 to 700 nanometres (white light) to unequal amounts, where rays of wavelength 650 nanometres (red) dominate?' This terminology is most helpful, but when, as throughout this book, I revert back to 'white' and 'brown', you will know what I really mean.

Now we are back to one of the original themes of this book – nature's colour palette. Remember the artist with violet pigment, blue pigment . . . red pigment – all colours resulting from pigments? Pigments are just one mechanism or factory for making colour. Nature's palette, I have asserted, boasts an array of colour factories, and each will be unveiled as we journey through the spectrum. This chapter, as it happens, is the pigment chapter. In that case, this is the chapter the artist may find least remarkable through familiarity except, perhaps, for one fact – this is the *ultraviolet* chapter. I will go on to explain, but currently we are stuck on brown. In any case, it would be helpful first to understand how pigments work, for the sake of comparison with nature's other colour factories if not to make art more interesting.

Clearly, it is the fur of the vole that makes it appear brown. Remove the fur and the vole has near-white skin. Although the main role of fur is to trap a layer of air and so provide a thermal barrier, another function is to control the colour of the animal as it would appear to an observer. The fur inherits this task because it is the outermost part of the animal. There are no adaptive requirements for the colour of vole skin because no light reaches it. This is best demonstrated in polar bears, whose skin is jet-black.

Detail of the colour factory

NanoCam is inserted into a single hair plucked from a vole's head. Passing through the cracks in the scaly outer surface, clumps of pigment molecules can be seen on the NanoCam monitor. Increasing the magnification on a pigment aggregation, one molecule is singled out. The pigment in this case is a 'melanin' – a type common for causing browns and blacks in nature. Melanins absorb most wavelengths in sunlight. But how?

The pigment molecule appears as a chain of atoms. In some parts of the chain the atoms are organised in a straight line, in other parts in circles. They are linked by 'bonds', or the sharing of electrons between orbitals of neighbouring atoms. Sometimes there are single bonds (where one electron is shared), sometimes double (where two electrons are shared); in fact the bonds alternate between single and double as one travels along the molecule. Many other animal pigments have this type of arrangement also, such as carotenes.

A short pulse of white light strikes the vole hair. NanoCam is positioned to observe the pigment at work. The electromagnetic rays mostly

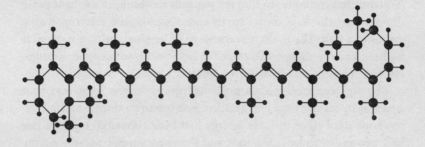

Figure 2.2   The carotene molecule. Large dots represent carbon atoms; small dots hydrogen atoms. Note the alternate arrangement of single and double bonds across the centre.

pass through the transparent outer part of the hair, as if it were a window. All in the same direction, they hurtle towards the pigment molecule until . . . the moment of truth. The whole molecule is shaken as rays collide with atoms. For some reason, NanoCam is overheating, but manages to remain focused on the event. First, as a light ray struck one of the atoms, an electron in its outermost shell became dislodged. And now electrons appear disturbed in nearly every atom of the molecule – the light ray has triggered a chain reaction.

Electrons are seen whizzing around the molecule, from atom to atom, travelling via alternate single and double bonds at extraordinary speed. The pigment is buzzing, almost alive.

What triggered this event was the passage of energy from the light ray to an atom of the molecule. This energy was enough to oust an electron from its residential orbit around the atom. The movement of successive electrons in the chain represents the passage of that energy – remember the First Law of Thermodynamics: 'Energy can be neither created nor destroyed.' So in theory, the electrons should circumvent the molecule forever. Well, they would if they were not losing a little of their energy package during each inter-atom leap. The energy lost is in the form of heat. That's why NanoCam began to overheat. That's why a black T-shirt, packed with pigments, will feel warm on a hot, sunny day.

NanoCam continues to film the pigment molecule as the light pulse illuminating the hair comes to an end. Quickly, the electron traffic grinds to a halt. The molecule returns to its original, resting state, as if nothing had ever happened, while the heat begins to dissipate through the hair. The optical effect is over. But what was it, precisely?

The pigment molecule extinguished many of the light rays that struck it by seizing their energy. Electrons began to shuffle around the molecule until all of the rays' energy had been converted to heat. The heat proceeded to leave the hair, and the energy input became energy output . . . until the next light ray collided with the pigment molecule and here we go again. So as long as rays continue to bombard the pigment molecule, the molecule will continue to steal their energy and dump it. The pigment molecule effectively destroys light. But do rays of all wavelengths share the same fate? Well, no, they don't, and as a result we have pigments in nature, and in paints, that appear different colours.

It takes a specific package of energy to dislodge an electron from an atom in the first place. This can be a very precise amount, or a range. The energy contained within rays of light varies, and is dependent on their wavelengths. Shorter wavelength rays, that appear blue, contain the most energy. If a pigment molecule requires precisely this energy to cause its electrons to dislodge, it will not be affected by green or red waves. Putting it another way, the pigment molecule will destroy blue rays but not those of green and red. Rays that are not destroyed by the pigment molecule are often reflected – spread out equally and, accordingly, thinly into all directions around the molecule. From here we can interpret how all of nature's pigments work (although variations on this system of electron movement do exist).

Back to the story

Different pigments feed on different-sized energy packages. Each type will consume rays of certain wavelengths in white light, while reflecting others. Those reflected are ripe for vision. The brown melanin in the vole's hair removes most of the violet, blue, green, yellow and orange rays from white light, and reflects most of the red rays. But the very small amounts of non-red colours reflected impart some visual effect, serving to provide an unsaturated red colour, which we interpret as 'brown'. As more of the non-red colours are reflected, the overall colour becomes pink. With equal amounts of all colours, of course, we would be back to white.

An important point about the colour effect of a pigment is its directionality − a pigment can only produce a reflection that is the same in all directions. The colour effect will appear the same from every angle or position. The misfortune of this, at least for the artist, is that a pigment's colour can never be the brightest found in nature. The energy in the reflected light is spread thinly over 360 degrees. Unless pressed

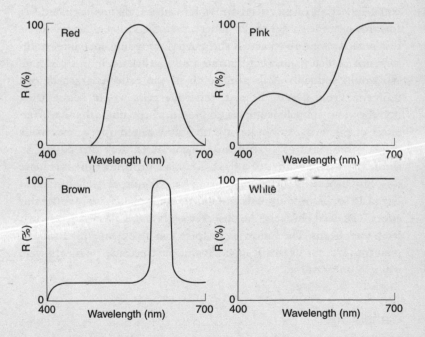
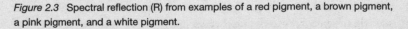

Figure 2.3 Spectral reflection (R) from examples of a red pigment, a brown pigment, a pink pigment, and a white pigment.

right up against a pigment, eyes are doomed to collect only a very narrow sector of the energy. And a small amount of energy equates to a dull light – it causes those cone cells in the retina to fire infrequently.

When we view a painting in an art gallery, we are unaware of the miniature factories within the paint that mass-produce colour. Every molecule in every pigment on the canvas continuously grabs rays from the white light bulbs above them, which sets electrons in motion. Electrons whiz around furiously in front of our very eyes, but we are oblivious to this energetic nano-world. We admire only the composition of (macro) slabs of paints.

The pigments in paintings reflect their rejected rays into every direction around a gallery, so that they appear the same colours from every corner of the room. And again, they appear relatively dull because our eyes grab just a small segment of that reflection. If a colour device could reflect all of its rejected rays into just a narrow segment, say towards just one corner of the gallery, then the visual effect would be different. As we walked across the gallery, we would encounter nothing until suddenly those concentrated rays struck us. In this position, we would suddenly grab almost all of the reflected rays at one moment. From a state of rest, the cone cells would begin firing impulses like a machine-gun, as opposed to the unhurried revolver-effect of pigments. Consequently the paint would appear extremely bright – the effect as we suddenly encountered and then passed by those rays would be that of a flash. But alas, pigments cannot achieve this. Pity, especially for Monet during his painting of game birds that day in 1879, because animals and indeed some plants *can* produce this effect. The dead pheasants he observed on his table flashed green light from their heads. The following chapter will investigate this further – pigments are not in this league. Again, pity, because pigments were Monet's only option.

Camouflage

With some understanding of pigments, nature's colour factory number one, we can progress to the next stage in the problem to be solved. That problem centres around voles that are perfectly camouflaged on motor-way verges.

I have mentioned that landscapes are packed with pigments. True. Leaves are generally green because of the chemical 'chlorophyll' in the 'chloroplasts' of their cells. Chloroplasts play host to photosynthesis, whereby the plant builds up its carbohydrate reserves. During photo-synthesis, water and carbon dioxide react to form glucose and oxygen. Light provides the energy to drive this reaction ... but not any old light. Only blue and red light. The energy in green light has no use to the plant, and so green light is rejected, or rather reflected. Like the rays not absorbed by any pigment, the green portion of sunlight is scattered into every direction, over 360°, while the blue and red portions are absorbed. Effectively, chlorophyll is a pigment.

A simple experiment can demonstrate the uselessness of green light to a green plant. Take three glass jars full of water, each containing an aquatic green plant. The first jar is lit by blue light only, the second jar by green light only, and the third jar by red light only. Soon, bubbles appear in the water of jars one and three, but not in jar two. The bub-bles are oxygen gas – one of the products of the photosynthetic reaction. So the occurrence of bubbles signals for photosynthesis, but under the blue and red lights only – green causes no reaction in the plant and is rejected.

As its food factory, chlorophyll is invaluable to a plant. As a conse-quence, the leaves of deciduous plants change colour before they are shed in the autumn (or at any time). They turn yellow, brown or red. The plant cannot afford to lose its chlorophyll, which is withdrawn from the leaves into the branches. Without chlorophyll, the leaf shows signs of its yellow, brown or red pigments, such as 'carotenoids', which are much cheaper to manufacture for the plant.

Dead leaves, along with twigs and soil, line a vole's runways within the undergrowth. Consequently, brown is a good, average camouflage colour for a vole. Point a spectrometer at the vole's micro-environ-ment and the reading matches that of the vole's fur rather well – a peak in the red region of the spectrum, combined with only traces of other colours. Repeat the NanoCam experiment with the human observer's eye fixed on the vole's micro-environment and the results are similar.

The same cone cells fire the same signals in roughly the same proportions – the brain interprets 'brown'. So view the vole within its natural
environment and all the cone cells fire similar impulses over the entire
human retina – red cones fire ten times more than both the green and
blue cones. The brain constructs a scene of brown; a uniformly brown
canvas. There is no colour contrast, and so no outline of a vole.
Ultimately, the human subject detects no vole. Camouflage has been
achieved. Seeing, as it is demonstrated by this case, is the detection of
colour changes, rather than the perception of objects. Only when there
is contrast between an animal and its surrounding environment can we
detect it *and* identify it. 'You see, but you do not observe' (Arthur
Conan Doyle, *The Adventures of Sherlock Holmes – Scandal in
Bohemia*, 1892).

This is all fact; there are no optical illusions. Also, our central problem of 'how kestrels manage to find voles that appear indistinguishable
from their background' is not a trick question. Vision is the principal
sense for most birds, and this certainly applies to kestrels – the kestrel
is not employing another sense, such as smell or hearing, to locate
voles. That would make motorway verges impossible feeding grounds,
with the overpowering influence of engine fumes and noise. So we
must hold up our hands – we are no nearer to solving our original
problem. But experiments involving NanoCam have been confined to
a human subject so far. Maybe our central problem can be solved only
by an experiment on the vision in question. Time to insert NanoCam
into the eye of a kestrel.

The kestrel's eye

With the kestrel relatively restrained within an enclosure for ease of
testing, NanoCam is positioned between four cone cells in the kestrel's
retina. Already there is the hint of something different – the *four* cone
cells are *all* slightly different. One contains large molecules in its cell
membrane that resemble those of human blue cones. Another contains
large molecules resembling those of human green cones; another has
large molecules like those of human red cones. But there is a fourth

type of cone. This 'fourth' cone looks similar to the other three, but the large molecules in its cell membrane, the molecules that react with light, are clearly different to anything in the human retina. Maybe the fourth cone holds the solution to our problem.

The kestrel in this trial is exposed to sunlight, and immediately the cone cells begin to fire impulses. The bird observes a whitewashed wall within the otherwise wire-fenced enclosure. Sunlight reflected from the wall careers towards the eye, traverses the cornea, is bent by the lens and . . . sparks begin to fly. All the light rays collide with the large molecules in the cone cells. NanoCam is shaken from every direction.

To its side, in one direction, NanoCam observes the large molecules in a red cone reacting to yellow rays (the so-called 'red' cones detect yellow and orange as well as red). To its side in another direction, a green cone reacts with the green rays. Again in another direction, a blue cone reacts with blue rays. All three cones fire electrical impulses through their cell bodies, which leap synapses and enter nerve cells. Eventually there are impulses racing towards the brain within the red, green and blue nerve fibres within the optical nerve. All impulses are equal, as one would expect from white light.

The cone cell to the west of NanoCam is also firing an electrical impulse. But no violet, blue, green, yellow, orange or red rays were observed to interact with its large molecules. The molecules underwent a transformation all the same. This is most irregular.

NanoCam has detected another type of light ray causing a reaction in a cone cell. This ray is a similar type of wave to those of red, green and blue, but its cycle length is shorter. The wavelength of this ray is 360 nanometres. Again, the impulses from the fourth cone cell flowed out of the cone, across a synapse, and into a separate, elongated 'optic nerve' cell via a short nerve cell. Within the optic nerve, this elongated nerve cell runs parallel with those servicing the other cones, and again relays the electrical impulse into a bulbous nerve cell within the visual cortex of the cerebrum. The brain interprets the impulse – it perceives it as 'ultraviolet'. *The kestrel is seeing another colour*, one beyond the repertoire of humans. The fourth cone cell is the 'ultraviolet' cone. The kestrel is a 'tetrachromat' – it has four visual pigments, each active in different regions of the spectrum.

There is nothing magical about ultraviolet. It is simply another colour in sunlight lying next to violet in the spectrum. The fact that we cannot see it makes it mysterious . . . but not magical.

Ultraviolet signals are fully integrated into the kestrel's colour vision system. Like red, green and blue, the *colour* ultraviolet is born in the brain. The image assembled (somehow) in the brain is made of red, green, blue and ultraviolet pixels. A rainbow appears wider to a kestrel than it does to us. Biologists examine the colours of animals and infer functions, taking into account the natural background hues. A small green animal sitting on a green leaf is likely to be camouflaged to a predator. But now we must question: 'Is that animal *really* "green"?' It is green to us, but is it green to whatever is looking at it? Add our invisible ultraviolet to the recipe and this is where we get a biological headache.

Ultraviolet – existence, history and measurement

The post-Victorian biologists became aware of ultraviolet's potential infiltration of the natural world when prototype cameras capable of ultraviolet photography appeared. For the first time, biologists' eyes were found wanting – they were not enough to answer questions on animal colour. Early twentieth-century biologists reacted by . . . doing little about it. Well, place yourself in their shoes when the news breaks that 'Now you all must consider another colour in your studies; only problem is, *you* can't see it.' It is difficult to deal with something that can't be detected. Nevertheless, hints that ultraviolet played a role in animal behaviour were soon dropped from odd studies, such as flies turning their bodies to face ultraviolet light sources, and flowers reflecting patterns of ultraviolet light. Now that the shock is over, biologists today certainly do something about ultraviolet. We have the equipment to measure ultraviolet reflection and vision, and the motivation to reveal an unknown world. And so we should do something about ultraviolet. Ultraviolet is equivalent to the sound from a dog whistle, which dogs hear but we don't. The ultrasound waves, nonetheless, are nothing special.

The spectrometer used to measure the vole's brown reflectance is known as the *Vis* type, where 'vis' denotes 'human visible'. There are other spectrometers, though, such as the *UV-vis* type, which extends the spectrum measured into the ultraviolet. I suggested that the fourth cone cell, which we now know as the ultraviolet receptor, might hold the solution to the original problem. At this stage the solution would seem rather predictable – surely the vole is camouflaged in the human visual region, but not in the ultraviolet, to which the kestrel is sensitive. This can be confirmed easily using the *UV-vis* spectrometer – a machine to add more to metaphysics, or the study of the first principles of nature and thought, than the seventeenth-century philosopher John Locke.

To demonstrate the range of the *UV-vis* spectrometer, it will be directed towards the whitewashed wall in the kestrel's enclosure. Sunlight illuminates the wall, and the spectrometer records whatever is reflected by the white paint. It emerges that white paint reflects everything that could be considered 'light'.

After passing through the molecules in the white paint, sunlight is reflected and enters the spectrometer. The diffraction grating divides up the spectral colours, sending the red light to the red sensor at the far end of the instrument, and blue light to the blue sensor in the middle. At the opposite end to the red sensor is an ultraviolet sensor, to measure the shortest waves of electromagnetic radiation. In this case the diffraction grating does send ultraviolet waves towards it.

The spectrometer records high levels of light between wavelengths of 300 and 700 nanometres – ultraviolet to red. The levels are high for the entire human visible spectrum and some of the ultraviolet (that lying next to violet in the spectrum, which I will term 'near ultraviolet'). The general peak in the sun's radiation output is the part that we and other animals see – this we call 'light'; the human visible spectrum plus near ultraviolet. Our eyes have evolved to see light precisely because of this – it is the strongest electromagnetic stimulus, the part of the sun's radiation that can be used most efficiently.

An interesting point is the effect of the Earth's atmosphere on sunlight. The atmosphere, particularly cloud cover, serves to absorb and 'scatter' sunlight, causing light levels to be reduced at the Earth's

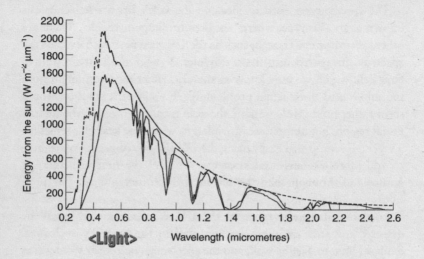

Figure 2.4 Sunlight intensity at the Earth's surface with various degrees of particles and water vapour in the atmosphere. The peak in intensity always lies in the range we term 'light'. Note that the wavelength units are in micrometres (x1000 nanometres).

surface. 'Scattering' involves random reflections, which average over all directions equally, and it will, incidentally, appear in a subsequent chapter of this book. But while on the subject of atmospheric effects, we can take a brief step back to the late Precambrian period, around 550 million years ago, when, according to planetary scientists, the Earth was covered in a blanket fog (see Chapter 10 of my previous book, *In the Blink of an Eye*). The foremost effect of cloud/fog cover is the one it has on ultraviolet light. Human visible light is somewhat reduced by cloud cover, but ultraviolet light is almost completely blocked. Consequently, the late Precambrian may have been an episode in the Earth's history that was nearly void of ultraviolet light. Soon, as the Cambrian period approached, around 545 million years ago, the blanket fog would withdraw. So the Precambrian–Cambrian boundary also represents a border between non-ultraviolet and ultra-violet conditions at the Earth's surface.

Curiously, the first eyes belonged to a trilobite (see Chapter 7 of *In the Blink of an Eye*), and included cornea-lenses that were transparent to ultraviolet light – they were made of quartz crystal. Did these eyes evolve in response to the strongest selection pressure, the introduction of ultraviolet light? Maybe these eyes saw ultraviolet light the best. Was ultraviolet light, therefore, the major selection pressure to drive the evolution of eyes in the first place? Unfortunately we will never answer these questions with certainty, because although the lenses of those first eyes exist today, the cone cells in the retinas do not. Trilobites became extinct some 280 million years ago, so we cannot test their response to ultraviolet light. We will never know if trilobites made use of the ultraviolet light that certainly became focused on to their retinas.

We use quartz lenses to record ultraviolet patterns of living animals. Glass absorbs ultraviolet light. The material of the cornea of our eyes absorbs ultraviolet similarly, and this is the reason for our lack of ultra-violet vision. In fact the blue cone cells in our retina *are* capable of detecting some (near) ultraviolet. So, although they would detect violet and blue light maximally, if our corneas were replaced with quartz replicas, our blue cones would offer some ultraviolet vision.

Camera lenses are generally made from glass, and so are not good at recording ultraviolet patterns (they are transparent to wavelengths just within the near ultraviolet region, such as 395 nanometres, causing professional photographers to use ultraviolet-block filters; that way nothing is photographed that they do not see and that could blight their composition). For a high price, crystal lenses are available for cameras. Like the trilobite lenses, these transmit not only ultraviolet, but also the complete human visible spectrum. The human visible spectrum, nonetheless, can be removed with an ultraviolet-pass filter, the reverse of the photographer's filter. Looking through a camera with an ultra-violet-pass filter one sees . . . nothing. Just black.

With a camera fitted with a quartz lens and an ultraviolet filter, and loaded with ultraviolet-sensitive film (ordinary camera film is not par-ticularly sensitive to ultraviolet light), an ultraviolet animal can be photographed under an ultraviolet light source. I first experimented with this apparatus on a spiny leaf insect in Australia – a type of stick insect covered in defensive spines. I wondered about a similar scenario to that we have postulated for the vole in its brown environment –

maybe the green camouflaged female insect was conspicuous in the ultraviolet, a region it can see itself. This would, in theory, provide a means of advertisement for potential mates.

A useful interlude – some known cases, and prevalence, of ultraviolet in nature

I had been given a single spiny leaf insect in a dry, ex-fish tank to look after for a few weeks. The body of the insect was almost as long as my hand. It was a bulky female and wingless – only the more streamlined and lightweight males have wings. Soon I began to wish that it was a male, because during my first day in charge the insect laid eggs – hundreds of seed-like eggs. Stick insects can reproduce without the influence of a male, although their offspring are exclusively female.

The hatching of the eggs staged a surprise. The juveniles were not miniature versions of their sluggish mother, but little machines scuttling around the glass tank like over-wound clockwork toys, mischievously searching for any sort of gap in the lid. Nor were they green. They were dark brown, with a red head. Their body appeared robust rather than fragile, but on closer inspection the illusion was exposed. Their long, thin, stick insect abdomens were curled up to offer a short, squat outline. Together with long legs to raise the body high off the ground, the overall body shape was that of a bull-ant. In fact the colour and frenetic movement of the juvenile leaf insect were also bull-ant-like. The juvenile of a camouflage icon, the slothful spiny leaf insect, was actually an ant mimic. In the wild, this strategy probably affords short-term protection as they disperse to new trees. Unlike bull-ants, fortunately, they did not possess large biting jaws or an ability to spray acid. Their visual appearance was entirely deceitful.

Insects pass through several moults during their lifetime. As the juvenile spiny leaf insects moulted for the first time, the stick insect style reappeared and remained throughout subsequent moults. But the growth of the insects posed a problem. Upon adulthood, they were certainly too big for their tank. So what to do with them?

Although I had once observed a crow eating an adult spiny leaf insect within Sydney, this insect was believed to inhabit more northern regions of Australia, so introducing the unwanted pets to the wild probably wasn't a good idea. Instead I gave most of them to the Australian Museum, where they became a living exhibit in a large enclosure. But I kept one adult female to test my idea of potential ultraviolet reflection.

There are several humane ways to kill an insect, and I opted for the freezing method. I left the insect overnight in a freezer. The following day, the insect was frozen solid, like an icicle. Obviously it must be dead. My equipment was ready, fixed in place on stands, although my camera flash lacked the power to provide suitable levels of near ultraviolet. Fortunately I found an old ultraviolet lamp in the Mineralogy storeroom, and directed it at the insect but away from the camera. I blew away the dust, focused the light on the insect's back, placed the ultraviolet-pass filter over the camera's crystal lens, and pressed the shutter button.

The camera was set to 'automatic', meaning that it had calculated the exposure time – the time the shutter would remain open and gather light for the photograph. The camera was detecting ultraviolet light only, and just that reflected from the insect's back. Some ultraviolet light must have been reflected – enough for the camera to calculate an exposure time. But, either because the reflectivity of the insect was low or the light source was dim, an exposure time of ten minutes was calculated.

I sat near the camera with its shutter open, as it proceeded to take the photograph. The long exposure time, I thought, would make up for the low ultraviolet levels, and over ten minutes enough light should be gathered to reveal any 'secret' ultraviolet pattern on the insect's body. The first five minutes were monotonous, but then something happened.

Somewhere around six or seven minutes into the photo session, the insect stood up and walked away! What looked like death was in fact latency. Insects contain anti-freeze chemical in their bodies, which prevents the formation of ice crystals that would rupture various body parts – they can withstand the freezing temperatures of winter nights (yes, even in Australia occasionally). But this event was a surprise all the same.

Take two. The insect was captured, killed instantly with ether fumes, and photographed successfully, for the full ten minutes. The result? It did indeed reflect ultraviolet from patches on its back. My colleague Moray Anderson, at Birmingham University, tested the vision of this species and found that, as expected, it could see in the ultraviolet. This seemingly camouflaged insect was in fact visible in its own secret world. It was perhaps communicating using private signals. Well, not quite the end of a tidy story, because the birds that feed on stick insects may also see the ultraviolet patterns. The only explanation I can offer for this seemingly counter-intuitive scenario is based on the penetrative power of ultraviolet light. The human visible colours reach further through the atmosphere than ultraviolet. Maybe, within the dense canopies of its environment, the ultraviolet stick-insect patterns are quickly absorbed by the pinball reflections from leaves and the surrounding air itself. Green, on the other hand, would reach much further, and even make it out of the canopy. In this case, bird predators, which hunt a range of insects, rarely get close enough to the stick insect to detect their ultraviolet patterns, which may serve as a 'private' signal to other spiny leaf insects at close quarters. This, of course, is just an idea and should be tested with experiments.

An alternative function for ultraviolet signalling appears likely in a bird-eating-mammal situation. Small black and white primates live among African canopies. The otherwise black Diana monkey, for instance, reveals a brilliant white throat. This primate is prey to large eagles, such as the harpy eagle, which like the kestrel possess ultraviolet vision. Not surprisingly my antennae began to twitch – are the white patches of these primates also ultraviolet patches? And would they be bright enough to reach the eyes of harpy eagles that hunt from within the forest canopies?

White is known as a 'broadband' colour, because it contains light rays of a broad range of wavelengths, or a range of colours. All the colours in the rainbow, in fact. It reflects much of what the sun casts upon it. Like the whitewashed walls of the kestrel enclosure, white matter may also reflect ultraviolet light – just another colour in sunlight.

The Monet painting *Poppies at Argenteuil* features an array of red spots, representing poppy flowers, with white clouds dominating the otherwise blue sky in the background. It is the red dots that immediately grab our attention. Aim a *UV-vis* spectrometer at a poppy flower and the reflection is revealed just as we see it – a red light, that's all. No ultraviolet. Now move the *UV-vis* spectrometer over a cloud and all is not as it seems. All the human visible colours, violet through to red, are joined by ultraviolet. Monet's white paint is reflecting all the colours that strike it. White pigments could be thought of as dormant. The energy contained in sunlight triggers no reaction in the white pigment. No electrons are encouraged to race around its molecules; no heat is produced. A white T-shirt, accordingly, does not feel warm on a summer's day. By definition, white pigments are not true pigments, in the sense that they remove no light waves from sunlight by seizing their energy via an electron effect. At best, they are borderline pigments, and a better definition will be explored in the Green chapter. Our concern here, though, is with the harpy eagle, which would clearly interpret *Poppies at Argenteuil* differently. The white clouds would appear more brilliantly white and overpower the effect of the poppies. Monet's accomplished composition would lie in ruins. But just as a Harpy Eagle would see more in *Poppies at Argenteuil* than us, do they also accentuate the white patches of small primates?

I joined up with Nick Mundy, a primatologist at Oxford University with an interest in vision research, and postgraduate student Nadia Khuzayim. Since the living primates are exceptionally active, and would certainly not hold a pose for minutes, dead specimens were needed. Nadia examined the fresh skins of several small black and white primates held at the natural history museum in Edinburgh, including the Diana monkey, and with great success. Ultraviolet photography of these skins exposed strong ultraviolet reflections from the white regions (with the exception of some yellowish-white regions). Our idea tested positive. But this experiment did not reveal the function of or consequences for the ultraviolet patches.

It is not just us but all primates that lack ultraviolet vision. We are all missing ultraviolet cone cells in our retinas. In other words, primates cannot see their own ultraviolet markings. Within the forest canopies, the primate ultraviolet signatures could only possibly affect bird competitors or predators. Maybe the ultraviolet reflections enhance the white flash effect as the Diana monkey reveals its chest. A sudden flash of light certainly can trigger a surprise or stunning effect in an observer, and may buy the primate a few valuable seconds to escape when pursued by an eagle. Many grasshoppers are known to deploy bright pink wings from beneath their camouflaged green covers, to shock any predator as it takes to the air to escape. So a shock tactic may be the purpose of the ultraviolet and white patches of small primates. Time to move on, however.

Apologies for my stick insect and monkey distractions, but there is, here, something relevant to our kestrel story. Private signalling is the principle I began to suggest for the voles. Do voles reflect signals in the ultraviolet only for the benefit of their own species, where kestrels are unwanted spectators? Or are the voles fitted with an ultraviolet shock response that, although to be deployed only as a last resort, may be leaked and detected by kestrels almost in an act of betrayal? Clearly we must first establish the ultraviolet status of voles by placing their hair under the ultraviolet photographic set-up, and by employing a *UV-vis* spectrometer.

Bad news. As the ultraviolet photographs of the vole hair are developed, no patterns emerge – the pictures are uniformly black. The *UV-vis* spectrometer reports no reflection of light below a wavelength of 400 nanometres either. The vole does not reflect ultraviolet light. A most inconvenient result, spelling the end to that nice, neat story and sending us back to the drawing-board . . . nearly. Actually we have made good progress.

The prevalence of ultraviolet in animals

Before attempting a new approach to the vole–kestrel problem, we should consider any pointers in the primate scenario above concerning the occurrence of ultraviolet coloration in animals. Namely, it would be interesting to know whether most white objects were also reflecting ultraviolet. That would give us a head start when considering species to study that may be more than meets the human eye. I have also considered violet pigments in a similar vein – since violet lies next to ultraviolet in the spectrum, so violet pigments may encroach on the ultraviolet range. I have experimented only twice with this idea, but with positive outcomes.

The violet snail is a predator that blows bubbles under which it floats at the ocean surface. Its shell is a typical snail-shape, and grape-sized. As its name suggests, it is violet – it contains violet pigment in its outer shell layer. But those violet pigments, unknown to the nineteenth- and twentieth-century biologists who first studied this animal, are also ultraviolet. The ultraviolet camera pictures the entire shell. A rather straightforward finding.

More interesting, perhaps, is the marlin's back. I photographed the back of a large striped marlin soon after capture to record the deep violet coloration. My first photographs were taken with an ordinary SLR camera, but then I switched to the ultraviolet camera set-up. Again the results were positive – the marlin's back emerged as both violet and ultraviolet. However, a closer inspection of the two cameras' photographs revealed something interesting. On one side of my composition I had captured several parasites – 'fish lice' (copepod crustaceans) that live on the skin of the marlin. The body of each fish louse was extremely flat and, from above, appeared of keyhole size and shape. The circular part of the keyhole shape was the head shield, covering most of the body parts and allowing only the simple eyes to protrude.

The fish lice were difficult to spot from a distance because of their camouflage. They were violet-coloured, and lying on the violet back of their marlin host. Fish lice are susceptible to predation by 'cleaner fish', because although marlins are known as the fastest swimming fish, for most of their lives they cruise slowly. Cleaner fish are even encouraged by fishes such as marlin, specifically to pick off their parasites and dead skin. Hence the violet camouflage colour of the fish lice – an evolutionary response to foil cleaner fish. But the interesting or surprising point is found in the ultraviolet photograph – the fish lice were difficult to spot in this, too. Both the marlin and the fish lice were reflecting ultraviolet similarly – camouflage occurs also in the ultraviolet. Recent studies suggest that cleaner fish are likely to possess ultraviolet vision,

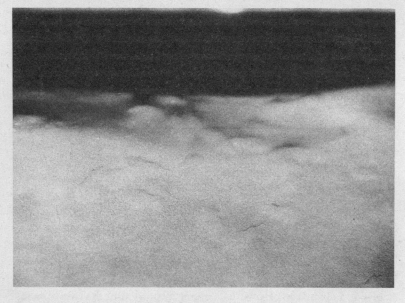

Figure 2.5 Fish lice (copepod crustaceans – oar-footed shrimps) on a marlin's back, in ultraviolet light only. The camouflage as evident to the human eye is perfect even in ultraviolet – the fish lice can barely be seen with artificial ultraviolet vision.

and ultraviolet light does penetrate at least the surface, marlin-cruising waters. Ultraviolet camouflage, as a result, is critical to fish lice survival.

Ultraviolet coloration is actually not uncommon in nature. It is now established that most insects and birds possess ultraviolet vision and as a consequence many also include ultraviolet in their coloration. The new cases of ultraviolet vision and reflection in reptiles, fishes and marine invertebrates are uncovered weekly, along with some nice stories of adaptation. The St Andrew's Cross spider of Australia, for instance, lacks ultraviolet vision but has evolved an ultraviolet-reflecting silk, which it spins in the centre of its web to attract flying insects that do see in the ultraviolet. The spider, meanwhile, can have no idea how it manages to attract so many insects to its web! Another case where ultraviolet is used positively to attract attention involves many birds, when chicks beg for their parents' food as they arrive at the nest. Ultraviolet happens to be the dominant colour from inside the gaping mouths of chicks belonging to at least several species, including the blackbird and house sparrow.

It is worth noting that in terms of vision and behavioural interactions we are concerned only with *near* ultraviolet; light of wavelength 350 to 400 nanometres (from where violet begins). This falls into a category of ultraviolet we term UV-A (wavelengths 315 to 400 nanometres). The ultraviolet blamed for causing skin cancers belongs to the UV-B category (wavelengths 280 to 315 nanometres). UV-B is harmful to animals because it damages the precious DNA and RNA, resulting in mutations, the death of cells and sometimes malignant transformation. This is why we wear sunscreen with ultraviolet protection . . . and so do some animals.

Although plants are better known for their natural sunscreens, many tropical reef animals also secrete solar protective chemicals. The saddle wrasse of Hawaiian reefs secretes sunscreen into the mucus that covers its skin, and is even able to vary the concentration depending on the level of ultraviolet content of the incoming rays.

The final category of ultraviolet, UV-C (wavelengths 100 to 280 nanometres), is the most damaging of all. Fortunately only 1 to 10 per cent of the sun's UV-B reaches the Earth's surface, and none of the UV-C. The Earth's upper atmosphere prevents UV-C infiltration, while the ozone layer also contributes to blocking UV-B (which is why the hole in the ozone layer is such an important issue). But in this book we are concerned only with colour – light that plays a role in animal interactions rather than inducing protective measures. Nonetheless, UV-C and UV-B serve to illustrate why animals have evolved colour and vision in the UV-A and human visual spectrum (wavelengths 350 to 700 nanometres) – this is the range of the sun's radiation that is always brightest at the Earth's surface.

With an understanding of what ultraviolet is, how it can play a part in animal interactions, and the knowledge that kestrels see ultraviolet when hunting for voles, we should attempt to wrap up this case. Ultraviolet does appear the main suspect in how the kestrel finds its prey, but the enigma remains that voles do not reflect ultraviolet. At this stage we are forced to leave theory completely behind and return to our experiments, taking no short cuts. This time we will test the complete natural system – *a kestrel hunting a vole in the field.*

The ultimate test

NanoCam is once more inserted into the eye of the kestrel, but this time the kestrel is released from its enclosure and allowed to enter its natural hunting grounds – fields bordering a motorway. Via a transmitter, NanoCam relays images back to a monitor, which we can observe.

The kestrel flies along a course parallel to, but around thirty metres from, the motorway. Its eyes look directly below and ignore the passing cars.

Electromagnetic waves with wavelengths around 650 and 520 nanometres reflect from the exposed soil and dead leaves, and fresh leaves, respectively, then pass through the kestrel's retinas as it surveys the land. Reactions take place repeatedly in the large molecules of the red and green cones, and impulses fire from these cones. The kestrel views a scene of browns and greens, but no vole. Suddenly something else happens – an ultraviolet cone cell fires an impulse. A train of impulses soon begins to fire. The kestrel changes its course, taking a left turn, and flies in a straight line towards a distant woodland and away from the motorway. The ultraviolet cones cease to fire.

The kestrel's response is to change course. It makes a 180° turn and heads back towards the motorway. Reaching the same point as before, the ultraviolet cone fires again – another train of impulses. The kestrel keeps flying, slowly. The impulses end, then start again, then end, then start again. The kestrel, now only a few metres from the motorway, takes a left turn, to once more travel parallel with the cars. This time, however, the ultraviolet cone fires a relentless train of impulses.

At this point we would not expect to find a vole beneath the kestrel because voles reflect no ultraviolet light. And we are right – there is no vole there, yet the kestrel finds this particular part of the field very interesting. It now begins to fly slowly over the same area backwards and forwards, ignoring the rest of the field. So what, exactly, holds the kestrel's interest here? The answer – chemicals on the leaves and ground that reflect ultraviolet light only. Ultraviolet pigments can exist inde-pendent of living matter. More importantly, in this case, is where those pigments came from.

Voles are territorial. They always use the same runways, which act like roads within their habitats. They mark their runways with chemi-cals secreted from specialised glands and conveyed by urine. These chemicals are signals, not only for the navigational benefit to their producer, but also to relay information about territory, social rank and mating potential to other voles. Some of the chemicals secreted within their runways are pheromones – they target the sense of smell. But is there also a chemical that has targeted vision?

NanoCam is positioned within the chemicals deposited by a vole in its trail, in this case on the surface of a dead leaf. Different molecules are imaged on the monitor, along with a strong incoming beam of light in the distance – the sun has moved from behind a cloud. As the light rays strike the molecules, most – the pheromone molecules for instance – remain motionless, but one stands out as different. This molecule looks different, certainly, in that it has a unique arrangement of atoms. But this is also the only molecule to react with the light rays.

The reaction is different to that in the brown pigment molecules. Only the shortest waves of sunlight – the ultraviolet waves – are reflected. But rather than being absorbed, the waves of all the other colours in sunlight appear to pass by the molecule, unaltered in their path. No doubt that this molecule is an ultraviolet pigment, although it is transparent to all other colours. The rays of those other colours con-tinue through the vole urine and are stopped in their tracks only by carotene pigments in the dead leaves below (which would absorb ultra-violet rays, along with violet, blue and green rays). It is the light rays that are rejected by the urine pigment molecules and enter the atmos-phere that are the important ones in this case. They are ultraviolet rays and may be detected by voles *and* kestrels.

Although early mammals generally lost ultraviolet vision, it was either maintained within one group, or re-evolved within that group as mammals dared to leave their undercover lifestyles after the dinosaurs had gone. Today's rodents are descendants of that group. They can see in the ultraviolet.

Now we know that voles not only leave trails of pheromones but also ultraviolet pigments, both of which they can detect. Ultraviolet is not reflected by the leaves and soil of the vole's background, making ultraviolet pigments stand out. Unfortunately for the vole, they stand out also to kestrels. Finally we have our answer.

The solution

Because kestrels cannot easily detect the well-camouflaged voles themselves, they look out for vole trails. They scan the land-type favoured by voles, such as motorway verges, and switch to ultraviolet mode – they ignore everything but the ultraviolet pigments that mark vole trails. When an ultraviolet cone cell fires a signal in the retina, a vole trail has been found. The more frequent the ultraviolet signals, the more vole trails in that area, which tells the kestrel that this is a good place to hunt. If we look out of an aeroplane window while flying low over land we see occasional roads running through the countryside. But these all concentrate on a village or town, so the density of roads reflects the density of human habitation. As the kestrel flew over the field, it followed the increasing density of vole trails and stopped at the equivalent of a vole village. The trails were simply a tip-off.

Now the kestrel can forget about ultraviolet and switch to movement mode – it will continue its hunting based on vole movement, which only works on the camouflaged prey because the kestrel knows where to look and because it has an additional string to its visual bow.

There is a region in the kestrel's retina that provides its most sensitive vision. This appears as a simple depression or pit in the retina filled with the same liquid ('vitreous humour') that fills the eyeball. It acts as a negative lens because the material of the vitreous humour has a lower refractive index than the retina, forming a 'telephoto' component. The image focused by the cornea and the main lens is shifted backwards by the negative lens, giving a longer overall focal length and a magnified image in that small region of the retina. This telephoto region is fixed on vole village, and NanoCam is manoeuvred into it.

The kestrel hunts for a brown vole against a brownish background. At this point we can redescribe the role of our, and the kestrel's, eyes as seeing colours, not seeing objects or shapes (this definition will be fleshed out as the book progresses). Outlines of objects on our retinas may be immaterial if the colour contrast is poor. The kestrel does not pick out the outline of a vole from its background. While it may be looking directly at the vole, it may not know so if the background is a similar colour – the retina forms a picture of brown, brown and more brown. This will be explored further in the Green chapter. But still, at a lower level, there may be *slight* differences in the shade of brown between the vole and its background, albeit not enough to distinguish the vole's shape. Since the kestrel's head is viewing the same small area over a short time period (as the kestrel continues to glide slowly), the individual cone cells are unchanged in the signals they fire.

Now a vole passes through the kestrel's area of view. Some cone cells *do* vary the signals they fire, although only slightly. Where one group of cones is focused on a pale brown leaf, as the vole passes over that leaf the signal changes from pale brown to dark brown. This gives the kestrel something to detect – movement. The kestrel's telephoto component helps to make the slight signal significant – the signal becomes a wave of dark brown flowing across the kestrel's retina. Movement in the field becomes translated to variations in cone cell firings. Variations *can* be perceived. Although the shape of the vole is not distinguished at any point, the wave is detected, and the kestrel must hope that 'wave' equals 'vole' The kestrel halts its slow glide and breaks into a hover. It assesses the situation. Is that variation in cone firing really down to a vole? After a few seconds it decides 'yes'.

The kestrel folds in its wings, dives to the ground and deploys its talons while it continues to track the wave front. Slam! Talons are finally sunk into vole flesh. The hunt is over. The kestrel, thanks to its ultraviolet vision, emerges the victor.

Fortunately the vole can rest on its high reproductive rate – its birth rate and death rate are in harmony. Voles as species survive . . . at least for today. Who knows what technology tomorrow's mutations may bring to this particular war . . .

A tonic for Darwin

Here is the first lesson that the eye is not perfect – it is not a *universal* visual organ. In this case, I refer to the human eye that cannot see ultraviolet light. There is a part of the world around us that we simply do not see. There are butterflies flashing warnings, birds showing off costumes and spider webs lighting up among grasses – all in the ultra-violet. These visual spectacles take place in front of our very eyes, yet we are completely oblivious to them. We simply cannot detect part of the world where there is so much activity, which also includes the shin-ing trails of voles. Indeed, that is why we have the problem posed at the beginning of this chapter.

At the beginning I mentioned that the solution could never be fully conceived. True. How can we pretend that an extra colour is there in the outside world, a colour we can't see? How should we react when we are told that a white Australian crab spider, which sits apparently 'totally camouflaged' on white daisy flowers, is actually extremely con-spicuous due to its ultraviolet colour? We will probably never envisage the seventh colour of the spectrum, because *the human eye is not per-fect after all*. On this account, *Darwin had no cause to worry*.

So colour factory number one on nature's palette is 'pigment', where nature and art truly cross paths. Indeed, melanin pigments from the ink sac of the cuttlefish *Sepia* were employed directly to prepare the artist's sepia paint. But pigments are perhaps the learning stage in under-standing nature's colours, because ultraviolet aside, they spring the fewest surprises. The workings of the pigment machine may be a little complicated, but its visual effect is rather too predictable. I studied pig-ments part-time until my colour addiction escalated. Soon I began to seek out the hard stuff. The next chapter will cover nature's most tech-nically brutal but at the same time inspired colour factory – structural colours.

COLOUR 2

violet

The problem:
Why is the harmless Malayan Eggfly butterfly so spectacularly conspicuous, when its predators are birds with good vision?

During my journey in 1989 through the rainforests of North Thailand that cloak the colossal, ever-rolling hills, a flash of violet light memorably distracted from the green. Green is everywhere, from the picture-book moment when the mist curtain rises in the early morning to the seriously competitive red of sunset. Other than green there are only colour voids – black or at best brown 'holes' in the backdrop where gaps between leaves fail to be filled by more leaves behind. Sunlight making it through the gaps meets only dark, absorbing pigments beneath. These pigments may lie within the bark of trees, but surely some must belong to the primates that can be heard rustling through the canopies or howling out songs. Unfortunately they cannot be seen. Occasionally a bird will pass into view, while there is only the thought of tigers. The violet flash, nonetheless, was obvious – the butterfly responsible almost landed on my head. It left a lasting impression amidst the green.

The insect accountable for the violet flash may have been a female Malayan Eggfly butterfly.

Alfred Russel Wallace was first to discover the Malayan Eggfly (*Hypolimnas anomala*) in 1869, during his expedition through South East Asia. It is of average butterfly shape and size and the male of the species is uniformly pale brown with a row of insignificant white spots fringing the wings. The spots are barely visible, leaving an overall drab façade. The female, conversely, is attractive. Its wing shape is indistinguishable from that of the male. And its colour . . . well, actually its wing colour is predominantly black, although an elegant, velvety black, while retaining the row of tiny white spots and adding another parallel row of fewer but larger white dots, further from the wing edge. But then there is the violet.

The violet coloration is not always visible – it only appears when the wings are held at certain positions. But when it is visible, it is striking. Out of the luxuriant blackness, as viewed when the wings are near closed, a violet glow materializes on each forewing. The glows swell as the wings are opened sleepily. First a violet spot near the furthest edge of each wing, then the bleeding of violet into the surrounding blackness until the wings are fully opened and violet dominates the forewings. The hind wings remain black. The effect is reversed as the wings are slowly closed once more. The opening and closing of the wings is accompanied by an undulation of violet – at full flying speed the effect is a pulsating series of hypnotic violet flashes.

The brown of the male Malayan Eggfly butterfly can be explained in one word – camouflage. This butterfly species has no defence against predatory birds, so camouflage would appear a rational tactic. So why has the female of the species evolved such vivid coloration? Why should it attract attention? This appears blatantly counter-intuitive and will be investigated in this chapter, beginning with the cause of the spectacular violet.

Nature has a near monopoly on this metallic-like, coloured effect; rarely is such a dazzling and dynamic colour encountered in the art world. I have, nonetheless, encountered one useful comparison.

A classical exception

Villa Borghese in Rome was home to the Borghese family, whose collective passion was art. Over the centuries, they amassed many classical oil paintings and marble sculptures by artists such as Caravaggio and

Bernini, and helped to preserve these masterpieces, so making a significant contribution to the history of art. Even the walls and ceilings of
the villa are swathed in classical frescos, while Roman-style mosaics
divide the rare marble floors. Light reflected from pigments in the paint
and stone strike the eye from every direction, in every room.

As explained in the previous chapter, just as landscape views fade
into the distance, paintings fade with distance too – both suffer the
shortcomings of pigments in that they dilute their reflected rays equally
into all directions. Our eyes, then, can only ever grab a tiny segment of
that reflection, and so most of a painting's colour disappears into the
air around us. If a colour device could reflect all of its rejected rays into
just a narrow segment, say towards just one corner of a gallery, then the
effect would be different. As we walked across the gallery, we would
encounter nothing until suddenly those concentrated rays struck us –
imagine walking into a laser beam (a highly concentrated beam of light
which does not splay out). In this position, standing between the painting and the corner of the gallery, we would suddenly grab almost all of
the reflected rays in one go. From a relaxed state, the cone cells would
begin firing impulses relentlessly. As a result the paint would appear
vibrant – the effect as we suddenly encounter and then pass by those
rays would be a flash.

In one small, corridor-like room of Villa Borghese, dominated by its
rich red background, a flash of light *is* encountered. This room exhibits
the 'micro-mosaic' pictures in the collection – scenes constructed
(painstakingly) from tiny coloured stones rather than brush-strokes of
paint. During a quick scan of this relatively less spectacular room, one
mosaic quite literally caught my eye.

Orfeo, a mosaic of 1618 by Marchello Provenzalle, depicts Orpheus
with a violin and various animals at his feet. It was two mallard ducks
that attracted my attention. Their green heads indicated their gender –
they were males. Females have brown bodies and heads. This is a
sexual trait found also in pheasants – the pheasants on Monet's table in
Vétheuil in 1879, with green heads, were also males. The male pheasant and mallard heads attract females, but not just because of their
colour, which contrasts well with the rest of the body, but also because
of their brightness. These bird-heads burn with green flames.

Monet failed to capture the dynamism of this green. As he paced around his subject matter, the iridescent green flare leapt around the pheasants' heads. First the crown lit up, then the throat. A wave of iridescence flowed over the feathers as Monet proceeded to walk, but this was not reproduced on his canvas. The green on the canvas stayed where it was as Monet viewed it from different directions. As I have mentioned, Monet gave us a hint of something extraordinary by painting the head black with a single green streak. The streak of green did contrast well against the black and so drew the eye more than any other colours in the picture. But the brightness effect – that dazzling, metallic sheen we see from compact discs or holograms on credit cards – was absent. The ducks in the mosaic *Orfeo*, on the other hand, literally leapt from the composition.

Most stones in the mosaic achieve their colour through pigment effects. Electrons jump between orbitals of the minerals' molecules when struck by white light, absorbing some wavelengths and rejecting others into all directions in the process. The green stones selected to occupy the ducks' heads also contain pigments – green pigments – which absorb the rays of all other colours in white light. But this time the pigments do not reflect the green rays but allow them to pass directly through the molecules, unaltered in their paths, like green-bottle glass.

The green colour seen from *Orfeo*, nonetheless, *is* reflected from the stones. The reflection is all about the physical shape of the green-pigmented stones. They are faceted, like diamonds. So as the green rays strike the angled rear edge of the stone, rather than exiting the stone they reflect from it at (for instance) right angles, as if the rear edge was a mirror. They reflect back into the stone towards the other side of the diamond shape, and from there they reflect back out through the stone and into the atmosphere. The green rays exit the stone in the direction from where they came – in a single direction, forming a beam. 'Beam' is the all-important character of this visual effect, in contrast with the splayed-out reflection from ordinary pigments. Accordingly, Provenzalle was able to capture the 'life' in the colour of the mallard's heads – he possessed 'structurally coloured' stones (although pigments did play a filtering role, it was the shape of the stone *structure* that caused the beam-type reflection). Without the green pigment, chandeliers and cut-glass 'crystal' tableware possess a similar sparkle. This is the optical effect to be explored in this chapter. Alfred, Lord Tennyson, made a useful comparison between a structurally coloured emerald and pigmented grass, which reflect the same rays. 'A livelier emerald twinkles in the grass,' he remarked.

A 'soap bubble' on the wing

Like *Orfeo*, the violet effect of the Malayan Eggfly butterfly is clearly not like anything usually encountered within an artist's frame. It does not obey the rules of pigments – the rays leaving the wings of this butterfly must be packed together in a narrow beam. But where do these rays go when the violet disappears to our eyes as the wings flap? And, come to that, what happens to those colours in sunlight that are not seen?

Butterfly wings are flat, membranous plates supported by thickened veins, analogous to leaves on that level. A magnifying glass will reveal microscopic scales lying almost flat against the surface, appearing like overlapping tiles on a roof. The scales are generally rectangular and attached to the wings by short stalks. Their stalk sides are nearest the body and their free ends directed towards the wing edges. In the case of the female Malayan Eggfly butterfly there are two layers of scales in the violet region, as evident where scales in the upper layer have fallen off. The upper layer consists of 'cover' scales, while the lower layer is made up of 'basal' scales.

Observing the cover scales under an ordinary light microscope, parallel stripes emerge out of the blur on each scale surface. Clearly there is microscopic structure to the scales – fuel for an *electron* microscopic examination. Scanning electron microscopes employ electron beams rather than light beams. This means they can image structures at extremely high magnifications; magnifications where reflected light rays are no longer resolved or focused.

To prepare a butterfly scale for the scanning electron microscope, it

Figure 3.1   Scanning electron micrograph of several scales on a butterfly's wing. The white scale bar represents $\frac{1}{10}$ millimetre.

is first sliced in half to reveal the internal structure, and then coated in a one-atom-thick layer of gold. At this thickness the gold will not modify the structure of the scale, but will allow electrons to be reflected from its surface and to be imaged in a 'micrograph' – the electron equivalent of a photograph. In 2002 Akira Saito, an engineering scientist from Osaka University in Japan, selected a cover scale from the violet region of a female Malayan Eggfly butterfly and prepared it in this way.

In the electron microscope, the single scale, just a hundredth of a millimetre wide, was quickly imaged at 3,000 times magnification so that it filled the viewing screen. The parallel stripes were unmistakable, running along the length of the scale, and were seen to be jutting out from the scale's surface. They will be known as 'struts'. The struts were linked by cross-ribs, whose function is to provide stability to the whole structure, with nothing between them – only a 'window' to the basal scales below. Increasing the magnification further, to 50,000 times, more structures show up. On either side of the struts are minuscule 'ridges', projecting sideways. Certainly the scale that at first appeared flat is distinctly three-dimensional.

The structure of a single scale is akin to a microscopic library room, without the books. The flat base of the scale represents the library

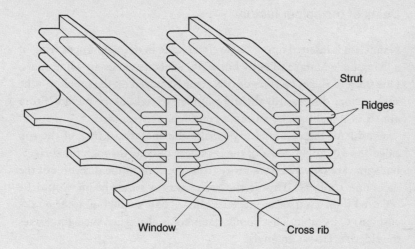

Figure 3.2 Diagrammatic nano-structure of a female Malayan Eggfly butterfly cover scale, showing struts, cross ribs, windows and ridges. Each ridge is about 100 nanometres thick; this complete section is fifty times smaller than the width of a human hair. The struts can be seen as fine parallel lines in Figure 3.1.

floor, the struts represent the columns of bookcases, and the nano-ridges represent the individual shelves. The shelves, in this case, slope – they are parallel with each other but not with the floor. They form a stack five shelves deep.

It is the shelves, or rather the 'ridges', that could be important to the optical effect – they are absent in the cover scales of the *male* Malayan Eggfly butterfly that lacked the violet colour. They are missing also from the *basal* scales in both males and females. The basal scales in the female are packed with granules of a melanin pigment, while the male basal scales reveal a more sparse distribution of a slightly different melanin. The melanin was the cause of the black colour in females and brown colour in males (like the melanin in vole fur). No other pigment was found in this species – the material of the scales is clear and transparent. At this point one should return to the *cover* scales and examine their effect on light waves.

Detail of the colour factory

NanoCam is inserted into a single cover scale in the right upper wing of a living male Malayan Eggfly butterfly, while its wings are wide apart. One stack of ridges on the side of a strut is imaged on the monitor – the shelves slope downwards slightly from left to right of the picture. From above, at a slight angle, a beam of sunlight arrives.

Crash! All the coloured rays strike the smooth surface of the top ridge at a 90° angle. Most of the rays pass directly through the surface, but some are reflected back away from the scale in the direction of the incoming sunlight. This is the same way a torch beam would be reflected from a mirror or indeed a billiard ball would bounce from the cushions of its table. That comparison holds also when the light strikes the ridge surface at an angle.

So why have the sun's rays divided into two paths? The reason, simply, is down to refractive index differences. Air, we have learnt, has a refractive index of one. The material of butterfly scales is the protein chitin, and this has a refractive index of around 1.56. As refractive indices go, that's a sizeable (although not enormous) difference. So as light passes from air to chitin, it effectively 'recognises' a boundary at the interface. Some light rays will pass through the boundary, others will reflect from it. The greater the difference in refractive indices, the more rays that will take the reflection route. In this case the difference is 0.56, which translates to about 5 per cent reflection and 95 per cent transmission. Five per cent of the sun's rays are reflected from the upper surface of the top ridge. But the top ridge also has a lower surface.

NanoCam follows the 95 per cent of the rays as they travel through the chitin of the top ridge, and nothing happens until they meet the lower surface. Here is another boundary for light – an interface between chitin and air – and the events of the upper surface are repeated. NanoCam records the second crash as 5 per cent of the remaining rays are reflected from the lower surface, in the same direction as those from the upper surface. Again, the rest of the rays pass through the boundary, this time into the air 'layer' below the top ridge.

NanoCam takes a moment to image the two sets of reflected rays, which, since they are following the same path, superimpose. Of all the rays, however, only those of a very specific wavelength superimpose exactly – their wave profiles match precisely, or are *in phase*, and cannot be separated on the NanoCam monitor. The rays of all other wavelengths are *out of phase* – their crests do not overlap. The 'in phase' condition is known as constructive interference; 'out of phase' as destructive interference. Only waves that constructively interfere live on, and in this case they have a wavelength of 450 nanometres – that's the wavelength for violet.

The violet light rays leaving the butterfly scale continue their path through the atmosphere. Any animal whose eyes cross over that path will capture a strong violet flash. The source of that violet flash I first saw in the rainforests of Thailand may have been revealed.

The same passage of light rays happens in the wall of a soap bubble or a layer of oil on water – both are thin layers of transparent materials. They have upper and lower boundaries and a different refractive index to their environments (air and water). Their thicknesses are just right to cause light rays reflected from their upper and lower boundaries to be in phase.

The colours are said to be 'iridescent' or 'metallic' because the eye perceives them to be bright. I mentioned that about 5 per cent of the sunlight is reflected from the upper surface of the scale's top ridge, and they are joined by another (approximately) 5 per cent from the lower surface, making the reflection 10 per cent of the incoming light. A pigment spreads its reflected rays over a complete hemisphere and so the small eye will collect *less than 1 per cent* of them. So this butterfly scale will appear more than ten times brighter than a pigment when its reflection comes into view. Again, in animals, this type of coloured effect is known as *structural colour*, because it originates from the physical nature of a structure as opposed to the electron activity within a chemical pigment.

So far I have described NanoCam's images from the top ridge on the strut. It is clear now that the ridges are responsible for the flash of colour – they are the reflector. And a single ridge can reflect 10 per cent of the violet rays in sunlight – a strong reflection when compared to the pigmented colours we are most used to. But there is more to this reflector – it hasn't finished yet.

NanoCam moves slightly to focus on the second ridge. The 90 per cent of the original sunlight that has passed through the top ridge soon encounters the second ridge, where it meets another boundary. Again there is a crash. The upper surface of the second ridge reflects another 5 per cent of the sunlight, and in the same direction as the two reflected rays from the top ridge. Again the violet rays are in phase, except now there are three rays leaving the scale, providing around 15 per cent reflection. The effect of a *stack* of ridges is becoming clear.

The lower surface of the second ridge reflects another 5 per cent of what is left and so on for the upper and lower surfaces of the third, fourth and fifth ridges. By the time the lower surface of the fifth ridge has played its part, 50 per cent of the violet rays have been reflected,

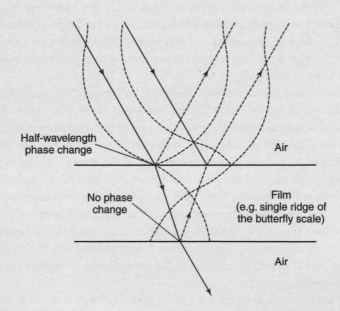

Figure 3.3 Schematic diagram of thin-film reflection, such as that from a single ridge of the butterfly scale when rays of light strike it at an angle. The direction of the rays (straight line) and their wave profiles (curved lines) are illustrated. Incoming rays are indicated by solid lines; reflected rays by broken lines. Note that the two reflected rays constructively interfere.

while the remaining 50 per cent, and the rays of all other colours, pass beyond the stack of ridges and into the basal scale below. A 50 per cent reflection would appear most vivid, and certainly would explain that violet blaze in the rainforest.

The sunlight that struck the scale came from one direction only – above. Now the butterfly is reorientated so that sunlight approaches its wings at a 45° angle from the left-hand side. NanoCam remains positioned within the cover scale, and observes the incoming beam of sunlight until the same top ridge is struck.

This time the light rays crashed into the upper surface of the ridge at an angle. Other than that, there was not much difference. Some rays reflected from the upper surface, others continued their path through the stack of layers, during which some were reflected from ridges deeper within the stack. The reflected rays head not into the incoming sunlight but away from the ridge at an angle, just as a billiard ball would bounce from a cushion. This time, however, the violet rays did not superimpose exactly – the peaks and ridges of each ray reflected did not overlap. In fact they destructively interfered, and were no longer reflected. The female Malayan Eggfly butterfly does not appear violet when sunlight strikes its wings from this angle. Some rays, nonetheless, did constructively interfere.

The same principle of 'multilayer reflection' still applies, only this time it is the ultraviolet rays that are reflected. The butterfly is strongly coloured in the ultraviolet – a 'colour' it can see. The ultraviolet reflection, like the violet reflection, occurs in its own specific direction, this time at an angle towards the right-hand side of the butterfly. We, of course, see nothing, only the black pigment beneath the cover scale.

Back to the story

To summarise, violet is reflected when sunlight strikes the opened wings of the female Malayan Eggfly butterfly from directly above, at 90°. At glancing angles, ultraviolet is reflected, but there is no room for other colours in this reflector. The *Urania* moth, on the other hand, reflects a different part of the spectrum.

There is a spectral trend in these multilayer reflectors. Longer wavelengths are reflected at 90° to their surface, shorter wavelengths at more glancing angles. In the case of the female Malayan Eggfly butterfly, those longer wavelengths were violet, and the shorter wavelengths were ultraviolet. But increase the wavelength of the rays reflected at 90°, and most of the spectrum can be accommodated. Red can be reflected at the normal position, then as the angle becomes increasingly glancing we pass through orange, yellow, green and eventually blue. And this is possible when the thickness of the ridges, or 'layers', increases slightly. Thicker layers are indeed found in the spectacular *Urania* moth, which reflect a rainbow of metallic-like colours.

The under-appreciated outcome of the multilayer reflectors in the scales of butterflies and moths is that the colours change when the insects flap their wings. We are often consumed with the brilliant colours and patterns from butterflies as they are fixed in display cabinets, and as they appear in books, always with wings wide open. We forget or even fail to acknowledge that these may be only part of the colours and patterns that have evolved in these insects.

The living butterflies flap their wings and often attract mates in the air. Observer butterflies, consequently, are faced with a colour pattern that changes with time, perhaps repeating every half-a-second, in tune with the wing-flapping cycle. In other words, butterflies with metallic-like (structural) colours have evolved a visual signal that has a time element, comparable to a short video recording rather than a photograph. The observing butterflies receive a series of fixed patterns (the frames in the video recording) which serve as a combination lock. Only when all patterns have been received, at the right (flying) speed and in the right order, is the lock opened and the mating pattern recognised. A male identifies a female of its own kind with great precision. The butterflies make no mistakes with such an intricate mating signal. There are no embarrassing attempts to mate with the wrong species. The point of this discussion on flight is that those pictures we see of brightly coloured butterflies are only a *part* of the colour story.

Here I have been referring only to the metallic, structural colours of butterflies, not pigments. Pigments are common in butterflies too – even more so in moths. In fact the combination of structural and pigmented colours adds more to the visual signal as the butterfly flaps its wings. Not only does the colour change, but the pattern also. Pigmented circles can be drowned out by structurally coloured squares at certain stages in a flapping cycle. Remember, pigments will appear the same from every position in the flapping cycle.

On the shoulders of giants

Some 150 years before Wallace discovered the Malayan Eggfly butterfly, Isaac Newton sat in his study at Woolthorpe, Cambridgeshire, contemplating the peacock feather. Newton had made great steps through his double-prism experiments and understanding that the sun consists of a spectrum of colours and nothing more. The prismatic effect, in fact, was a case of a structural colour – it is the physical structure of the prism that shreds sunlight into its spectrum of colours. Maybe, he thought, the peacock feather was another case.

Newton had taken a special interest in the work of glassmakers; in particular the thin slivers of glass that accidentally pile up on the workshop floor. He was captivated by the iridescent effect of the glass slivers – glass was famed for its transparency, yet here it was more intensely coloured than any of the dyes *intended* to provide a hue to his clothes. In the dimly lit kiln works, Newton's clothes faded into invisibility while the glass slivers on the stone floor made full use of the little light available and even sparkled against their murky background. Newton had seen this effect before.

Peacocks had been introduced to the gardens of the British aristocracy from India, and their feathers quickly found fame. Newton could not fail to notice the iridescent effect of the 'eyes' at the ends of the male's tail feathers. He could not examine the internal make-up of the peacock feathers, but the slivers of glass provided the same visual effect and so, he thought, held the answer to the peacocks' iridescence. In the case of the glass, the stack of thin slivers had obviously caused the iridescence. Maybe there was a similar stack of thin layers within the peacock feather, Newton postulated. In his book *Optiks* of 1704, he did point out that some form of optical interference was responsible for the peacock's splendour. But he was unaware of the wave nature of light and accordingly the overlapping of peaks and troughs in the reflected wave profiles that achieve constructive interference. Nonetheless, the effect of multiple layers, or interfaces, on light had entered scientific thought.

It was not until 1923 that this science progressed, when the chemical engineer Clyde Mason noted with more precision the similarity of the peacock's iridescence with a stack of thin films. By this time the multilayer reflector had been fully explained in physics, and Mason could even predict the thickness of the layers that surely must have lain within the feathers. Yet he could not observe them.

At around the same time, but completely independently, the eminent Indian scientist C.V. Raman pondered colour in nature from his desk in Bangalore. Raman became famous for his application of physics to a number of subjects, particularly gemmology and the type of stone encountered in the micro-mosaic *Orfeo*, but his work on iridescent bird feathers rather bypassed the scientific world. Unfortunately he published only one scientific paper on the subject of animal iridescence – unfortunate because what he deduced was right, and could have injected some pace into the subject.

The birds of India – the Indian jay, the Himalayan pheasant, and, of course, the peacock – fascinated Raman. He was able to deduce multilayer reflectors with the sort of clarification that did not appear until some half a century later. Fortunately he was able to convey his fascination to large audiences during dozens of lectures entitled 'Birds, Beetles and Butterflies'. These were the days when colour slides were not yet in vogue. But Raman was able to paint word pictures of, for instance, the Indian jay, so graphically that the listener could see in his mind's eye the succession of coloured bands of alternating deep blue and greenish blue changing dramatically when the bird was in flight; dull and drab when the light is from behind and turning a brilliant green with metallic lustre when lit from the front. It was clear that to Raman birds were beautiful and a subject fit for serious optical study.

Soon after Mason's educated estimation and Raman's inspired genius, the scanning electron microscope was invented and turned the study of materials on its head. By the mid-nineteenth century, this machine moved from the Physics departments of universities into the Biology buildings. Instantly, the biologically aware Thomas Anderson and Glenn Richards took a giant leap in magnification and were able to view the struts, ribs and even the ridges of butterfly scales. This shook the study of invertebrates in particular. Those state-of-the-art drawings revealing the hairs on fleas with some wonder, from just the previous year, were instantly obsolete. Now bacteria on those hairs were coming into focus.

Quickly the optics of butterfly scales were calculated and general butterfly iridescence was explained – 'butterflies contain multilayer reflectors'. In 1962, H. Durrer subjected a peacock feather to electron microscopic examination, and found almost what Newton had imagined – minute granules, a quarter of a wavelength of light in size, arranged regularly. Although not thin layers, stacks of granules on the same scale were believed to approximate a multilayer reflector. As good an explanation as it seemed, further investigation of this structure lay ahead.

The next major exponents of the subject were the British biologists Eric Denton and Mike Land. In an impressive series of papers, they revealed a whole range of multilayer reflectors in a diversity of animals, from marine snail shells to cat's eyes. They were responsible for adding 'mirrors' to the biological literature, a subject that will be called upon in the Red chapter. More appropriate to this chapter is the work, beginning in the early 1970s, of the US biologist Helen Ghiradella. Helen revealed variations in the architecture of different butterfly scales, which she matched with their precise coloured effects, including ultra-violet signatures. Immediately physicists joined in and used butterflies as examples of the optical phenomena that had otherwise failed to attract public attention. Physicists continue this work today in various labs around the world, where example after example of wonderfully precise optical devices are discovered on a monthly basis. The hunt is on for a completely novel type of optical device in a butterfly scale that could be employed commercially, but that's a story in its own right.

I was lucky enough to stumble into this subject of structural colours in nature in 1990, during a zoological tour of Australia. I couldn't help but notice the abundance of bright animal colours, as compared to their scarcity in the more sun-deprived Britain. Not only were the pigments generally brighter – reds, blues and yellows as opposed to browns and greys (with exceptions of course) – but shimmering, iridescent colours were almost the norm. Most beetles were living jewels and reef fishes appeared to glow. My personal favourite was the Magnificent Riffle bird, where the male's iridescent throat, like the wings of the female Malayan Eggfly butterfly, was accentuated by the stateliness of its velvety-black, feathered surround. When the throat feathers caught the sun's rays they changed from violet to green with the brilliance of highly polished metal. I worked, however, on tiny marine animals such as seed-shrimps (which will appear in the following chapter), where, in 1993, I found diffraction gratings. Diffraction gratings are microscopic, corrugated surfaces, and are another type of structure that causes colour, as found on compact discs or in holograms. This finding rounded off the suite of simple optical reflectors known in animals.

Everything was done and dusted with respect to the *major* cause of iridescent colours in animals . . . until the turn of the twenty-first century. Then another category was added to the list – photonic crystals. Here things really get complicated. To fully explain how photonic crystals work we must draw on a modern branch of physics known as 'quantum optics' or 'photonics', where all the probabilities for light ray diversions are considered, rather than just their average path through a structure. Photonics concerns the control of light in the manner that electronics depends on electrons.

Although biologists had been observing photonic crystals without their knowing, the first photonic crystal to be identified as such in an animal belonged to the sea mouse *Aphrodita* (a marine, polychaete worm). *Aphrodita* has the appearance of an iridescent mouse, due to its coat of spectacularly iridescent hairs and spines. My Australian colleagues Ross McPhedran, David McKenzie, Lindsay Botten, Nicolae-Alexandru Nicorovici and I found that these hairs and spines are filled with 'nano-tubes' – tubes smaller than the wavelength of light, which collectively prevent certain light waves travelling in certain directions. So although they act essentially like multilayer reflectors, where the walls of the nano-tubes substitute for layers, they can additionally control the colour of light reflected with some precision by preventing the reflection of some

Figure 3.4 A scanning electron micrograph of the two-dimensional 'photonic crystal
fibre' of the sea mouse *Aphrodita* sp. – a section through the wall of a spine. The
white scale bar represents 8,000 nanometres

wavelengths. The identification of a photonic crystal in nature attracted
more physicists to the subject of natural structural colours, and in 2003
three more photonic crystals were recognised in animals.

The first recognition resulted from a conversation with a then under-
graduate student, Dominique Driver, who had expressed an interest in
iridescence in nature and needed an animal to work on. One tends to
notice 'interesting' displays of light in animals that are somehow 'different', and this becomes a good excuse for electron microscopic
examination. I showed Dominique my collection of 'visually interesting' insects, and from between a spectacular butterfly and a giant,
gregarious beetle, she picked out a humble weevil, rather modest in its

visual appearance. The diminutive specimen was essentially black – although this time an unimpressive, run-of-the-mill black – with a littering of blue-green iridescent spots. The spots, nonetheless, were unusual in that despite their not uncommon metallic effect they appeared the same colour from every direction. Most metallic colours change as the eye moves around them.

The spots were in fact scales, just like those of butterfly wings although here they sprouted from the weevil's back. Dominique scraped off some scales, sliced some in half to reveal their insides, gave them a thin coating of gold and examined them in the scanning electron microscope.

The first observation was that the scales were round and flat. But as the magnification increased, something emerged within the scale. Zooming in further, tiny spheres, half the wavelength of light in size, could be seen packed within the otherwise hollow scale. They were packed regularly and tightly. I had seen this structure before. This was opal, which was well known, but always in gemstones. Here it was in an animal! Opal is a photonic crystal, and the technological repercussions of this find have only just begun. A fortunate discovery, certainly.

The second photonic crystal revealed in the same year emerged in peacock feathers. The peacock feather had not only commenced the study of animal structural colours – 300 years later it returned at its forefront.

In 2002 Shinya Yoshioka and Shuichi Kinoshita of Osaka University in Japan reported that the previously considered reflecting layers in the

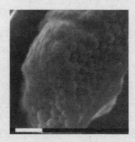

Figure 3.5   Scanning electron micrograph of the opal-like, three-dimensional photonic crystal of the weevil *Metapocyrtus* sp. (formerly thought to be *Pachyrhynchus* sp.) – a section within a scale. The white scale bar represents 1000 nanometres.

peacock feather barbules – the 'twigs' that come off the 'branches' of a feather – may actually be part of a 'photonic' system where quantum optics should be called up. Then in 2003 Jian Zi and his colleagues at Fudan University in Shanghai confirmed this idea using complex optical computer modelling, after being lured to peacocks in the wilds of southern China. So Newton made an unfortunate choice of models in the peacock; the eyes in its tail feathers are photonic crystals, whose elucidation demands computers, and large computers at that.

Because its details are demanding, I will not dwell on the theory of photonic crystals. My colleague Victoria Welch is in the midst of uncovering more of them in animals . . . the list is likely to expand quickly. But I will end their involvement in this book with the third photonic crystal to be revealed in an animal in 2003, which brings us back to butterflies.

The costs of bearing structural colours

A group of Hungarian biologists, chemists and material scientists, along with their Belgian physicist colleague, Jean-Pol Vigneron, paid close attention to certain populations of (lycaenid) butterflies. Typically, in most species the males and females exhibit different colours. The males often possess bright, structural colours on their upper wing surfaces, while females are brown; echoes of the pheasant and mallard scenarios. Darwin used 'sexual selection' to explain such male–female differences, where male ornamentation was selected through female preference for extreme traits. Today we think differently. Generally, ornamental traits including bright colours are thought to be honest signals of individual quality, although still driven by sexual selection – females look for the fittest males rather than a particular fashion.

The Hungarian–Belgian group targeted a butterfly species called *Polyommatus marcidus* – a 'Mountain Blue'. This species lives at different altitudes in Iran, along the slopes of the Elbrus mountains. This study began by using male and female individuals caught near the base of the mountains, whose upper wing surfaces were scrutinised. In the electron microscope the male wing scales appeared different in shape to

Figure 3.6   A butterfly photonic crystal lattice. This structure has long been associated with butterflies, but only recently identified as a photonic crystal.

those of the female. As the magnification increased, another difference came into view – male scales were endowed with photonic crystals.

Beneath the 'struts' of a male scale, visible through the 'windows' and broken edges of the scale, was a complex, three-dimensional lattice. An array of spherical holes were stacked up within the material of the scale, like miniature Swiss cheese but with holes of all the same size and spacing – in fact more like the opal structure but with the spaces and scale material swapped around in the form of a 'negative'. This lattice structure was missing from the female scales. It was undoubtedly a photonic crystal, and when passed through a computer model, it was identified as a violet-reflecting photonic crystal. The cause of the male's violet colour had been found. The female's brown melanin pigments were identified too.

More individuals of the Mountain Blue were examined, although this time from high altitudes, from regions above two and a half kilometres within the Elbrus mountain range. There were plenty of brown specimens to examine, but a complete lack of violet butterflies. Were there no male butterflies at such heights?

The species of these high-altitude butterflies was confirmed as *Polyommatus marcidus*, the same as those from low altitude, and their sexes were identified. Within the brown assembly were both males and females – the males here were brown too. Promptly the brown scales of males were examined in the electron microscope, and as could be expected they were without photonic crystals.

So what was going on here? Why did the male butterflies of the same species appear so different at different heights on the mountains?

Earlier in this chapter I mentioned that Australia was a good place to spot structurally coloured animals. One reason why Australian animals can afford to be cavalier with iridescence is that they inhabit a warm climate. They do not generally need to absorb heat – usually the problem is getting rid of it. As we now know, in the case of structural colours some of the sun's rays are reflected while the rest pass through the structure, so no effort is made to capture the energy within sunlight. Pigments, remember, involve the movement of electrons within molecules, and any rays they do not reflect or allow to pass by are absorbed. Absorption involves the conversion of light energy to . . . heat energy. In other words pigments can be good thermal devices; structural reflectors are not. And the fewer colours reflected by a pigment, the more energy in sunlight that can be converted into heat – black pigments are the best thermal device of all (remember the black T-shirt on a summer's day).

Now it makes sense why the male butterflies are violet at low altitudes and brown at high altitudes. At the base of a mountain the air is warm. The male butterfly can obtain more than enough heat, and even afford to lose the potential heat energy within the sun's violet rays. So it has evolved a mating signal that makes use of a violet flash, to which the female is receptive. The stronger the violet coloration, the more efficient the reflector – this is an honest signal of the individual's fitness. To maintain a perfect photonic crystal lattice in a scale, the host butterfly must be healthy – if unhealthy, the lattice may be the first object in its body to buckle. It must also be efficient at absorbing and utilising the energy from coloured rays other than violet to conceitedly expel the high-energy violet rays.

Ultraviolet aside, violet light contains the most energy in sunlight.

High up on the mountain the air is cold. To survive in these conditions, the butterflies must grab all the energy that is available to them and convert it to heat. Here the butterflies exist at the threshold of life and death. Reflecting violet light and its energy within would push them over the threshold. The butterflies high up on the mountains have evolved more brown, light-absorbing pigments and lost their photonic crystals. As for mating signals, they must place more emphasis on their pheromone chemicals and less on visual display. At least they stay warm.

The Malayan Eggfly butterfly, on the other hand, is not restricted within its environment by temperature, and the female possesses the more classical multilayer reflector rather than a full-blown photonic crystal. Still, the multilayer reflectors of this butterfly are as complex as they can be, and from a technological perspective are enviable for their delicate latticework at such a minute size. This structure is beyond the current limits of human nano-engineering, as advanced and astounding as that may be.

The Blue Crow butterfly

Although the Mountain Blue butterfly showed a good deal of colour variation within its species, a flick through a picture book of butterflies will also reveal the opposite scenario – not only can colour be remarkably stable within individuals of a species, but also different species can appear the same colour. Maybe, even, I should return to that violet flash in the forests of Thailand and double-check my butterfly identification.

The 'Thailand violet', to now thought to be the consequence of a Malayan Eggfly butterfly, could equally have derived from a Striped Blue Crow butterfly (*Euploea mulciber*). This butterfly was discovered some 100 years before the Malayan Eggfly, but shares the same geography, and general wing size and shape. Both butterflies are prevalent throughout at least South East Asia.

The male and female of the Striped Blue Crow butterfly appear visually different. The female exhibits the same form of intense violet patches on its upper wings, which come and go as the wings flap, but the remainder of the wings appears pale brown, with several obvious white stripes in the positions of the wing veins. The male Striped Blue Crow butterfly, on the other hand, closely resembles the female Malayan Eggfly butterfly. Beyond its violet zone, rich, black scales swathe its wings. This highly contrasting black background causes the violet to appear slightly brighter than it does in the female of the species, and occupies a slightly larger area of the wings. In terms of natural selection, it is the male who is at greater risk from visually-hunting birds, while the egg-bearing females are a little less conspicuous. Since males can mate with more than one female, the species would benefit from greater *female* survivorship than male. Such sexual selection and its associated colour trend were expressed also in the pheasant, mallard duck and Mountain Blue butterfly cases – it is common in animals.

Considering it is the main theme of this chapter, maybe we should investigate just how closely the violet colour of the male Striped Blue Crow butterfly resembles that of the female Malayan Eggfly butterfly.

A cover scale from the violet patch of a male Striped Blue Crow butterfly is coated with gold and placed in a scanning electron microscope. As the magnification increases, the struts are first to emerge out of the blurred image. They are spaced exactly the same as those of the female Malayan Eggfly counterparts. Also indistinguishable are the cross-ribs, the next structures to materialise in the microscope image. Finally, nano-ridges are evident – the component making up the reflector in the female Malayan Eggfly. And again there is a perfect match – the minute, intricate nano-ridges of the male Striped Blue Crow butterfly are identical in size, shape and the precise angle at which they slope to the nano-ridges of the female Malayan Eggfly. This match is something quite extraordinary. Even with our great technological leaps that have made it possible to make *almost* anything using modern manufacturing methods, we cannot make this structure. Yet this highly intricate optical device has emerged not just once but twice in animals, independently and in identical form. No problem.

It now seems that the female Malayan Eggfly and the male Striped Blue Crow butterflies produce identical colour displays – identical down to the wavelength of violet reflected, the angle at which it is reflected, and even the pattern variation and the change to ultraviolet as the wings flap. So which of these butterflies did I observe in Thailand in 1990? My only chance of distinguishing between the two possibilities would have been to observe the body shape of the butterfly, which unfortunately was impossible at the time.

Like all animals, butterflies are divided into groups based on their relatedness. Butterflies and moths comprise the insect group (or order) Lepidoptera. The highest level of relatedness, however, is the *species*. A species is, in the case of butterflies, a group of individuals that will mate exclusively with each other in their natural environment. *Hypolimnas anomala* (the Malayan Eggfly butterfly) and *Euploea mulciber* (the Striped Blue Crow butterfly) are different species. They would not mate with each other in the wild. There was never any gene transfer for precise, violet scales – those scales really did result from independent evolutionary paths.

The level of relatedness below species is the genus, a collection of related species; the next level down is the subfamily, a collection of related genera and species. The Malayan Eggfly belongs to the subfamily Nymphalinae, while the Striped Blue Crow belongs to the subfamily Danainae. It is their bodies that give away their relationships. The shape of the male and female body of the Malayan Eggfly is short and stout, while the Striped Blue Crow butterfly's body is always long and thin. The bodies of both species are an unremarkable brown in colour, but it is the shape only that is important to classification.

Maybe surprising here is that the Malayan Eggfly and Striped Blue Crow butterflies belong to different subfamilies. A little earlier I mentioned that 'a flick through a picture book of butterflies will also reveal . . . [that] *different species* can appear the *same colour*'. By this I assumed the 'different species' were closely related; sister-species even – those *most* closely related. But that is obviously not the case in this story. I will return to this point after finishing with classification.

Butterfly body shapes are governed by a large number of genes. The body is made up of a head, thorax and abdomen; the abdomen is further divided into segments. There are genes to code for the shape, length and width of each body part and segment, and also for the characteristics of the eyes, the three pairs of legs, a pair of antennae and the mouthparts. Many genes code for many, relatively large body parts. Colour, in contrast, is dependent upon few genes. Yes, even those sophisticated violet reflectors may evolve simply from the archetypal butterfly scale.

Evolution involves gene mutations. Mutations that code for changes in the body that provide a competitive advantage for the individual are retained within the species. Mutations with negative undertones are lost ... quickly – the recipient individual of such a misfortune is unlikely to survive for long enough to reproduce and pass on its inferior genes (or will appear unattractive). Because body shapes require comparatively many or major mutations to alter significantly, they are a more reliable indicator of butterfly relationships than are wing colours. This was not realised until the Victorian era, when the real science of zoology dawned.

Through the seventeenth and eighteenth centuries, intellectuals had amassed cabinets full of natural curiosities – bones, fossils, shells, feathers, insects – from the far corners of the Earth, although arranged in no particular order. These were passionately discussed, but conversations and arguments lacked an objective. A new and more systematic approach to nature was required in order to make sense of it all. Soon books became more widely available and, following the Linnaean system of classification, natural history collections began to take order. A coordinated effort to catalogue the world's wildlife had begun, albeit within only wealthy circles of society.

It was the nineteenth century, nonetheless, that became the great age of the natural historian. Here, the middle classes – doctors and clergymen – took over the butterfly nets. These collectors were rather more adventurous in their choice of locations and their rigour to achieve their goals, often risking their lives and enduring great hardship in pursuit of new, exotic species. Their collections formed the basis for the great natural history museums of the world, and even today the bulk of these museum's anthologies carry nineteenth-century labels. One of the most famous collectors of this Victorian era was Henry Bates.

The son of a manufacturing hosier, Henry Bates demonstrated an early interest in zoology. In 1848 he joined Alfred Russel Wallace on a momentous expedition to the Amazon, where he amassed a vast collection of animals and plants, of which about 8,000 were new species. But this was not the reason for why the expedition was 'momentous'. That lay in the colossal numbers of butterflies Bates had encountered. A painting by Arthur Twidle exists of Bates chasing a huge, spectacularly blue (*Morpho*) butterfly in Brazil, which captures both the tropical atmosphere and Bates's enthusiasm and determination to make a thorough, scientific collection. Bates, clad completely in beige linen with shirt-sleeves partly rolled up, and broad-rimmed hat, takes cover behind a sizeable tree trunk, amid random, winding vines and further shades of brown. His eyes fixate on the outstanding *Morpho* flying within a forest clearing, just beyond the curtain of vines, while his left hand clutches his rifle with unmindfully great might. One holds one's breath with the two natives in the painting, both armed with nets and crouching behind Bates, almost ready to leap from the frozen moment and capture that specimen before it's too late. 'Quickly,' one is tempted to shout, drawn to participate in the privileged opportunity. Needless to say butterfly collectors are no longer heroes today.

Bates had gathered a butterfly collection that included both whole insects and individual wings found unattached to bodies on the forest floor. The individual wings had been discarded – they indicated predation by birds, and so their frequency exposed the extent of predation. In a way it was those discarded wings that made this expedition momentous. Some species, it would seem, were preyed upon less, probably because they contained toxins. And that became part of a momentous story.

Henry Bates grouped together his butterflies based on body shapes, and instantly discovered a discrepancy in wing colours and patterns. Wing coloration, it would seem, had no bearing on evolutionary relationship – just as we could have predicted from gene numbers. As this was the 1850s, Bates would have been ignorant of genes, but that was not important for the discoveries he would make. Body shapes were reliable indicators of evolutionary relationships, and that's all Bates needed to work out to conclude that colour was purely adaptive ... and more. Bates proposed that some butterfly species were *mimicking* others.

The solution

As with the female Malayan Eggfly and the male Striped Blue Crow butterflies, Bates noticed the same colour pattern occurring in different parts of his carefully separated collection (carefully separated by relatedness). Some butterflies were known to contain chemicals that are toxic and distasteful to birds. Where two unrelated butterflies shared the same colour pattern, one would always contain toxins and the other would not. It was clear what was going on. Through trial and error individual birds would associate a specific colour pattern with toxicity, and in the future they would avoid butterflies with that pattern. With this a new niche opened up. Harmless butterflies that evolved the same colour pattern as toxic species would be attacked less by birds. Bates had uncovered mimicry. Today this strategy is known as Batesian mimicry, as a tribute to Bates's enterprise. There are also other forms of mimicry and some will emerge later in this book. And although Bates's discovery may seem undemanding, over 1,500 scientific papers on mimicry have been published since and are indicative of the importance of mimicry to evolutionary theory.

As one might guess by now, there is a most important difference between the Malayan Eggfly and the Striped Blue Crow butterflies – the Blue Crow is poisonous, the Eggfly is not. The Striped Blue Crow butterfly contains chemicals in its body that are poisonous and distasteful to birds. Its violet flash is a warning to birds, an honest signal of its toxicity. Both the male and female carry the toxins, and both the male and female generate violet flashes – a warning colour.

The Malayan Eggfly evolved after the Striped Blue Crow butterfly. Its immediate ancestor was without a violet flash, but genetic mutations occurred that coded for the violet multilayer reflectors in its scales, and suddenly (and of course without understanding why itself), the mutant lived longer. It was less prone to bird attack. So the harmless yet more conspicuous mutant avoided birds the best – a counter-intuitive case if it were not for the rules laid down in this environment by the Striped Blue Crow butterfly. The Blue Crow is known as the *model*; the Eggfly the *mimic*. *The problem at the beginning of this chapter has been solved.*

Still, although the evolution of the mimic butterfly makes good sense when evolution is considered, we should not casually sweep this case aside. We should take a moment to marvel at what has really happened here.

The first half of this chapter was near devoted to how the female Malayan Eggfly butterfly generated its colour. That was not an easy job. In fact the answer derived from 300 years of scientific study, involving some of the most eminent scientists. The structure causing the colour was amazingly intricate, and fine-tuned to cause an intense violet reflection that changed precisely in its pattern and alternation with ultraviolet as the wings flapped. The optics of its scales were excellent – today physicists strive to make such precise nano-structures that carry considerable information in their changing colour patterns for anti-counterfeiting purposes (the precise pattern can be measured to indicate authentication, but not easily copied). So although mimicry makes sense, and it is almost not surprising that a butterfly evolved to copy the violet warning colour, the physical copying of such an intricate, precise and sophisticated optical device is a visual self-tribute to evolution.

As for the evolutionary process of mimicry in this case, one must envisage that it took place step-wise in the female Malayan Eggfly butterfly and its predecessors. From a violet-less ancestor, the Malayan Eggfly predecessors must have begun with a weak violet reflection from few scales. Although far from a perfect mimic, birds may have avoided them if faced with a completely brown alternative individual – violet is part of the signal for toxicity. But only part. Nonetheless, the violet morphs within the population of this species would become more and more common through longer survivorship and increased mating frequency, consequently passing on the genes for violet scales, albeit any old violet scales.

The next morphs to appear and live on evolved increasing detail to their violet reflectors. The struts of the scales would have changed their spacing, and the nano-ridges increased in number from perhaps two to three. The next stage along this evolutionary road would see another ridge added . . . and so on until there were five. Further down the road the ridges would slope at a precise angle and become equal in thickness.

At this point the Blue Crow reflector had been completely matched. But along the entire evolutionary road, each stage of scale architecture provides a stronger and directionally precise violet reflection over its predecessor. This means that the more evolutionary advanced stages appear visually closer to the male Blue Crow reflector than do the less advanced stages. So birds generally targeted the less advanced stages over the more advanced stages – the genes coding for the closest scale-match at any time live on, making the next evolutionary step feasible. The process of evolution is theoretically viable.

Once the perfect match has been achieved, however, there are limits to the population sizes of the Eggfly and Blue Crow butterflies. Rises and falls in the numbers of individuals of one species inhabiting an area will directly affect the other species. This is because the bird predators must discover this warning colour through trial and error – this knowledge is not inherited, or something that is implanted in their brains at birth. They must *learn* to avoid violet flashes, and this can only happen if they are unrewarded or harmed when they prey upon a violet butterfly. But if there are more harmless than there are harmful violet butterflies, then any bird that attacks a violet butterfly for the first time is likely to select a harmless specimen. That bird now understands that there is food at the end of a violet signal.

If the same bird selects another violet butterfly, and it too is harmless, suddenly the mimicry system begins to break down. Even if the third butterfly selected is unpalatable, the bird is still likely to attack a further violet butterfly with the odds favouring food. In fact the maintenance of mimicry is all about probabilities. Simplistically we could assume there must be at least an equal number of models as mimics. This could well be broadly true, but certainly the minimum ratio of models to mimics for mimicry to work will be different in every case.

Finally we are left with just one loose end to tie up. The clouds have lifted on our seemingly unorthodox Malayan Eggfly colour scenario, where the female is conspicuously coloured and the male is camouflaged. As we have seen, usually sexual selection results in the opposite outcome, where it is the male's role to impress the females. But here the selection pressures for mimicry overcome those for sexual selection, explaining why the female Malayan Eggfly is conspicuous. What is not clear, however, is why the *male* of the Malayan Eggfly butterfly is *camouflaged*, and that question has yet to be answered.

Maybe it is the female who attracts the male in the Malayan Eggfly butterflies, in a role reversal. Then the female could exploit its bright colours also for mating purposes, where it is the male whose job it is to seek out the female.

One study by Japanese biologist Naota Ohsaki suggests, with good evidence, that in butterfly species where the males and females appear similar, birds attack the females more frequently than the males. If predation is biased towards females also in butterflies that show Batesian mimicry, as Naota's study also suggests, then the selection pressures on the females to evolve mimicry would be stronger than those on the males. However, quite how predatory birds distinguish between male and female butterflies that appear similar has not been resolved.

Maybe the evolution of the Malayan Eggfly subsequently affected the evolution of the Blue Crow butterfly. Certainly the Malayan Eggfly's appearance in the environment would have increased the death rate of the Blue Crow through a longer learning process of their toxicity for bird predators – the Malayan Eggflies dilute the toxic lesson. The Malayan Eggfly's appearance possibly became a selection pressure for the female Blue Crow to reduce its conspicuousness – its pale brown hind-wings do provide camouflage and its white stripes do serve to break up its butterfly outline (another subject to appear in a subsequent chapter). But this is all speculation, I'm afraid, so this loose end must remain untied for now. Rest assured, research in this area is ongoing.

A tonic for Darwin

This chapter holds the second lesson that the eye is not perfect. The birds that feed on butterflies in South East Asia can't distinguish between the body shapes of their prey when mesmerised by a strong-coloured wing pattern. Even though their potential prey is the most conspicuous animal in the whole forest in which they live, they choose not to eat them.

If the bird predators could track their prey butterflies using the violet flashes, but then make a distinction between palatable and unpalatable violet individuals, they would be better fed. All that is needed to make this distinction is a quick glance at the body shape – if it's short and fat it's food, if it's long and thin it should be left well alone. A seemingly simple task, but unfortunately the violet becomes just too distracting for the birds.

The bird's eye is drawn towards the violet flash as it moves, and away from the butterfly's body. This is a reflex action – it literally can't take its eyes off the alluring structural colour. Here, the bird's eye *is not perfect after all*. And this is not just a one-off – it is just one of many thousands of cases of mimicry where the eye is tricked through distraction. On account of the eye being too *perfect* to result from evolution, *Darwin had no cause to worry*.

With the hard-lined theories of physics now behind us, we can explore the more romantic side of colour in nature, beginning with a sight fit for a legend.

Although the subsequent chapters hold new ways in which animals appear coloured, the mechanisms behind these are rather simpler to explain than they have been to here. The stories I have selected to introduce them, nonetheless, *are* involved, with more than the odd twist in store. Next we will wait for the sun to set and watch the most 'enchanted' of nature's lights, still thought by some today as 'ancestral spirits'.

COLOUR 3

blue

The problem:
What really are the 'spirits of the sea' that left Philippine villagers seeing ghosts?

'Dear Dr Parker, I was advised by Asst. BFAR Director to consult you on the emission of bright light underwater in one part of our waters,' began a letter from the Bureau of Fisheries and Aquatic Resources, Republic of the Philippines, dated 5 August 1998. 'Although residents in the area observed it to be occasional, the last time it occurred was conspicuously long and bright that attracted a huge crowd to watch for it and the high-lighting by the media. A sincere desire to know what it is, what caused the luminosity ensued whereby academe in the area and BFAR institution were consulted and asked. No sampling was done but subsequent diving after the occurrence of the bright light did not give any idea, so that views on *superstition*, man-made, *ancestral* were referred to as the causes.'

The last line of this letter is interesting. Local people had observed such an abnormal radiance, like nothing they could associate with nature, that it seemed almost supernatural. They would have witnessed spectacularly iridescent beetles and butterflies; indeed tribal people in this part of the world once wove the dried insects into their most important costumes as a substitute for gemstones. They encounter birds of paradise that also seem to defy the laws of nature with feathers so elaborate and obviously useless for flight. They knew that nature was capable of almost anything ... except this blue glow. But what lay behind it? Even marine biologists emerged empty-handed.

There is nothing odd about the region of coastline in question here, at least during the day. A bay interrupts a long stretch of beach, and the sea within appears typical of this part of the world – an inviting blue-green in colour. But it does not glow. And the anonymous light is not simply an artefact of the night, where the darkness of space would provide the contrast needed to expose a steady radiance that loses out to sunlight during the day (a torch light appears bright at night but not in daylight, to make an analogy). No, the blue glow is absent most nights. But occasionally after sunset, when total darkness sets in thanks to a minimal moon and the absence of city lights, an intense blue 'flame' rises from the depths of the bay. Larva-like, glowing water quickly pours from the edges of this embryonic radiance and continues to permeate the dark sea until finally the whole bay is alight. And not just with a hint of a glow. This is something that can be seen from space.

Although the blue glow is not an artefact of the night, the darkness does aid its visual impression, just like the black surround of the violet butterfly's wings. The colour contrast is important but much more so the gap in brightness. A half-moon night witnesses minimal light, or near-zero intensity, so any luminous object will appear brightest against this background because there is almost nothing to dilute its effect. I say 'luminous' object. But without a light source such as the sun, how can pigments or structural colours shine? They reflect light, but at night (barring a little of the sun's reflection off the moon) there is no light source, and so nothing to reflect – pigments and structural colours do not work in darkness. Another form of colour or light display must exist on Earth.

Mushrooms

Hundreds of years earlier, the ancient Philippine people were rather more scared by a similar sight. Again, glowing lights were involved at night. This time, however, the lights glowed in a cemetery.

A section of forest had been cleared to make way for simple graves. The roots on the clearance perimeter had been severed, and the trunks attached had died. But *something* surely was alive. At night the trees glowed green, shining out for 100 metres or so, at least to the boundaries of the cemetery. Imagine emerging from the eeriness of a forest at night to observe a glowing green tree, and in a graveyard at that? Not surprisingly the people were frightened. They were recorded to have seen the 'ghosts of the dead'.

At night, during the First World War, life in the narrow trenches of France bordered on chaos as some soldiers manoeuvred potentially blindly around others attempting to prepare their ammunition. Disorder could easily have developed as the blasts of gunfire further raised adrenalin levels. But the soldiers possessed a more peaceful weapon that could help them to avoid fatal collisions. They owned glowing branches just like those in the Philippine graveyard.

The radiant wood was attached to their helmets and the ends of their rifles, and suddenly order was installed. Now they could easily see the trench traffic. If only the enemy forces could have viewed the trenches from above, they could have noted every movement of every troop even at night. But one war later, similar wood would favour the enemy.

During the Second World War, British paratroopers landed in the occupied territories of France. They were anticipating empty fields, at least that was the picture painted from Intelligence. It was night-time but it was not dark, not completely anyway. As the troops' parachutes opened they could see 'things' on the ground below – lights of some description. They were not expecting to see 'things'. They became nervous. Had their plans been leaked? Were there in fact enemy forces waiting below after all?

As the paratroopers approached the ground the yellowish green lights became brighter. Their hearts raced, right up until they landed. Then, nothing happened. No ambush and no explosion. The lights, they discovered, were natural. But they did not exist in Britain, to the misfortune of the architects of this landing, or rather the paratroopers making the landing who could not be forewarned.

During the same war, on the other side of the Earth, an American correspondent sat in his tent in New Guinea one evening writing his usual accounts of the days' events. He was still writing as the night set in, careful to practise economy with all his supplies, including torch batteries. Before retiring to his bed he wrote his usual letter home. The letter opened: 'Darling, I am writing to you tonight by the light of five mushrooms.'

Actually all these cases involved 'mushrooms'. The glowing trees in the graveyards were dead, like the wood in the trench soldiers' helmets, but they were infiltrated by living fungi. The fungi were responsible for the light. And the big brothers of these species, the mushrooms, are also known to glow.

In 1890 the Scottish explorer Henry Drummond entered the region of the Swan Valley in Australia. He came across a huge mushroom, weighing some two and a half kilograms, and out of curiosity he brought it to his hut, where he hung it beside the fireplace. As he approached his lodgings on return from a night walk in the 'bush' he witnessed a creepy image. What should have been the silhouette of his hut against a dimly moonlit woodland was rather a greenish yellow light beaming out of the windows of the otherwise shadowy building. As he flung open the door it was clearly the huge mushroom that glowed. It continued glowing for five nights until it dried completely.

Today many species of mushroom are known to glow in the dark, even where there is no light source whatsoever. The purpose of this is unclear. Some researchers suggest that the fungi gleam to attract insects that will assist in the dispersal of spores. But a very different question concerns the machinery behind the light. Obviously this light is neither a pigment effect nor a structural colour. The *precise* mechanism as it happens in fungi is currently unknown, but some clues are given in another historical account, this time from Scandinavia.

Above the Arctic Circle, the winter can experience perhaps just an hour or two of sunlight each day, so people here must work during the dark. In the past, pieces of wood infiltrated with a glowing fungus were common in this region and were used as lanterns to illuminate people's paths – a sensible practice. What is slightly worrisome, however, is that the same glowing branches were taken into barns filled with dry hay. Surely this was an accident waiting to happen? But nothing did happen. There was never a fire.

The lesson here is that although the fungus acted as a light bulb, it did not emit heat. Light bulbs turn most of their electrical energy – some 90 per cent – into heat. The light of mushrooms is cold. The explanation for this must involve chemistry.

Glow-worms

A colleague and friend at the Australian Museum, Peter Serov, knew all about my interest in colour, and in 1994 suggested an appropriate walk in his native forest. I had an inkling of what lay at the end of the path proposed, but really had no idea of the visual experience to expect. We would be heading to the little-known (at the time) Glow-worm Tunnel.

Peter picked me up along with other friends, including visiting professor Wim Vader (from Norway), from the Australian Museum in the centre of Sydney. It was early in the morning, but already warm, as the six-month-summer had begun. Peter owned an 'accidental' classic car, on the border of style-icon and old banger. I remain unsure about which side of the border it lay, but certainly this circa 1970 pale blue Holden seemed appropriate.

Old Holden cars were Australian-designed and built with the challenging bush and outback conditions in mind. They were tough, and today are popular not so much for their collectability as for their utility. Still, they possess an element of style against their sunny, palm-laden backdrop, in the manner of Cuban cars. All six of us climbed on to the two bench seats – front and back – and wound down all the windows. We welcomed the natural air-conditioning, although not the overwhelming noise. Nor the flies that were sucked into the car at pace, although ironically flies were the reason for this journey.

On leaving suburbia we joined a highway cutting through blue-green eucalyptus forests and a mountain range. As the land flattened once more we entered Peter's hometown of Lithgow, a million miles from the stress of the city. Beyond Lithgow, roads as I recognised them ended. Nevertheless, Peter's Holden made short work of the dirt tracks that carved through the woodland, its speed kicking up dust like a wagon train. We were glad of the large springs in our worn beige leather seats.

After dusting down we continued along a much narrower track on foot. My naïve reactions to the leeches that endlessly homed in on my feet caused some amusement. It was incredible first how they raced towards me from all directions and second how they managed to squeeze through the spaces between the threads of my socks before ballooning on blood. Nice. But really, whenever I stood still I observed them accelerating towards me along the leaves of trees and grasses, caterpillar-style. Intermittently I would remove first my socks, and then the leeches using salt, which causes them to vomit and fall off (again, nice). The reason for their prevalence is that we were following a stream – leeches need humid conditions.

Leeches aside, the stream was picturesque, bordered by moss-covered rocks and ferns. The tree canopy overhead was so full of holes that the sunlight leaking through, and reflecting from the water, demanded sunglasses. Our eyes had adjusted to bright light, and soon we would realise this.

At the end of the light lay a tunnel. This was the Glow-worm Tunnel, an old rail tunnel 400 metres in length, although the line was dismantled in the 1940s. The entrance prepares one for a cave rather than a tunnel. Ferns overhang the large hollow in the rock face, which is filled with darkness and nothing more. Several metres of path and walls are visible, but then it appears to end – at the back of the 'cave'. The ground inside is uneven and damp, and darkness falls with each step. Beyond those first several metres is complete darkness, at least to begin with.

The tunnel bends, so that once well inside neither entrance nor exit is visible. But quickly one's eyes adjust to the darkness. Pupils widen to allow more light to reach the retina, making the eye more sensitive to low levels of light. Then, after a few minutes, there appears more to the darkness.

From above and to the side, lights materialise. Minute, pinhead-sized lights against the pure black backdrop. Blue, glowing speckles. Thousands of them.

The feeling is that of camping in the bush, well away from the city lights, and observing the sky at night. There are minute stars every-where, packed in between the larger 'city stars'. Without interference from house and street lights, we see luminosity from deeper into space. And these pinhead stars appear just like the glowing dots in the Glow-worm Tunnel. But here we were standing in a tunnel, so what was causing these lights? If we could insert NanoCam into a single light, we would have an answer. And within the confines of this book, that is possible.

A single speckle of light is targeted and NanoCam is edged towards it. As expected, the light detected becomes increasingly brighter and the sensitivity of the monitor, adapted by default to the black background, must be adjusted coarsely to avoid a saturated image of blue on the screen. Now we see the blue light emerging from a teardrop-shaped object, unidentifiable at this stage. Could this be alive?

NanoCam has moved to the edge of the object, and now only blue is revealed on its viewing monitor. One final push and NanoCam breaks through the surface and into the centre of the teardrop. Little effort is needed, as the object is not solid but gelatinous; a sticky slime that glows. The magnification is increased to that where optical structures would be evident – they are absent. There is no structural colour. With further magnification the molecules of the teardrop manifest.

The molecules are mainly H_2O. Other molecules are interspersed within this sticky solution, but none distract from the overall trans-parency. There are no pigments. Even when NanoCam's spotlights are turned on, there is no reflection of any wavelength of light from within the teardrop, not even ultraviolet. So what is the source of the blue glow? The teardrop is beaming, but has no factory for light.

A hint of a clue is noticed. As NanoCam turns within the teardrop the light intensity fluctuates – during a complete rotation the monitor becomes gradually brighter then darker again, although always the same blue colour. Logically, NanoCam is turned to face the direction of the brightest light, and is driven straight ahead. The light continues to become slightly brighter as NanoCam sweeps aside the molecules in its path, until all is shaken by a sudden impact. There is something solid in the centre of the teardrop.

Again the solid object is transparent, but again it is glowing blue.

NanoCam moves left and right and the object vanishes from view. Move up and down, on the other hand, and the object is seen to continue. Evidently this is a 'thread', a transparent tube. It runs through the centre of the teardrop and appears to bring with it blue light. But from where?

NanoCam bores through the surface of the thread and manoeuvres upwards, from where the brightest light approaches. The thread is thin – the width of a hair – and bright inside. Even where it emerges from the top of the teardrop, it continues to glow – clearly the teardrop does not contain the factory for this light. Now the thread is exposed, but as NanoCam travels upwards towards the roof of the tunnel, it enters another teardrop.

The thread passes through several transparent teardrops, as if stringing the beads of a necklace, and all are glowing blue. But approaching the roof of the cave the teardrops cease. Here, NanoCam hits another wall.

This wall belongs to a further, much wider tube, a whole millimetre across and running in a different direction. It is made of the same material as the thread, as evident from both the force required to forge through its molecules and its transparency, although it is hollow. The large tube also glows. Yet even this is not the source of the blue light – no colour factories are present here either. It is, however, slightly brighter again.

Following the brightest light, this time within the large tube, NanoCam eventually encounters life. An object blocks the large tube and prevents NanoCam from advancing further. And this object moves. Here, at the edge of this life-form, the blue light is at its brightest.

NanoCam pierces the transparent skin ahead by squeezing between the outer layer of cells – animal cells – and enters one of four parallel rods inside the animal. The cells in the rod can be identified further – they are part of a kidney-like excretory system.

Running through the rod and around NanoCam are hundreds of minute, vessel-like tubules, branching to evenly infiltrate the entire rod. NanoCam passes through the wall of one of them. These are not blood vessels but are instead filled with oxygen, and travelling through them leads NanoCam all the way out of the animal via a large pore. The tubules are the branching ends of the trachea, the breathing system. This life-form can only be an arthropod – an insect- or millipede-like animal.

Tracing its path back through the tubules and into the rod, NanoCam tunnels into one of the commonest type of cells that pack the rod. These cells lie at the end of the trail of light. They glow feverishly.

Detail of the colour factory and a 'fibre optic' system

NanoCam focuses on a molecule of a compound known as a 'luciferin', and also on its neighbour known as a 'luciferase'. A molecule of oxygen, which originated from the tubules of the trachea, passes into view in the background and appears to be heading towards the luciferin and luciferase molecules. Soon, as anticipated, the oxygen molecule actually collides with the others and . . . there is a blinding flash of blue light.

Replaying the footage on the viewing monitor in slow motion, it is clear what happened. Energy is supplied by adenosine triphosphate (ATP) molecules (the energy-laden molecules common to all living cells), which hold the energy obtained from the animal's food. This energy fuelled the luciferin molecule to join with the luciferase molecule and form a molecular complex, which quickly reacted with the oxygen molecule to form an oxidised complex. This must have been highly unstable because immediately it broke down to form another molecule, known as an oxyluciferin, while the luciferase was released unaltered. During this decay, the energy that was absorbed to initiate the reaction was dumped, emitted as a package of . . . *light*. From out of nowhere within this cell came rays of light of a particular wavelength, in the blue region of our spectrum. They headed out through the cell wall and into the surrounding environment. First they passed through the animal's skin and into the hollow within the large transparent tube. From here they continued their travels, but in a controlled manner. Wherever the blue rays struck the wall of the tube they simply bounced back towards the centre, and continued to bounce throughout the tube's length. The tube acted as a fibre optic, where the light entering the tube was the same as that leaving it. Well, it would have been, were it not for the side branches.

The fine threads into which NanoCam first travelled are joined to the large tube. Light rays that strike the wall of the large tube precisely at a joint do not bounce off the wall as normal, but instead enter the thread. Although the thread is not hollow, light continues to flow along its length, bouncing off its outer surface and back into the centre of the thread. The reason for all of this rebounding of light is the refractive index difference.

Two materials are involved in these tube systems – a protein making up the tube walls, and the air that surrounds it and fills the larger tube. These have different refractive indices, so where light meets their interface, as we saw in the previous chapter, some is reflected and the rest passes through the boundary. Remember the butterfly example where around 5 per cent of the light is reflected? Well, in that case the light struck the boundary at a right angle. In the case of the tube, we have a contrasting situation – light strikes the walls at an extremely glancing angle. At such an angle *all* the light is reflected, bouncing along the length of the tube all the way to its end (or to a branching point for a thread) – this is how a fibre optic works. If the tube is bent sharply, then light can escape because suddenly the angle at which it strikes the interface between the protein and air becomes closer to 90°. Total internal reflection only occurs at extremely glancing angles.

As light passes through the finer threads that hang down from the roof of the cave, it also passes through the teardrops. The teardrop material has a refractive index close to that of the thread protein, so now the light can take two paths. It can continue through the thread or it can enter the teardrop. Light eventually enters the atmosphere from the teardrop – the teardrop glows.

Back to the story

The chemical reaction that took place in the animal's cells is known as *bioluminescence*. Bioluminescence was the cause of the blue lights in the otherwise darkness. The Malphigian tubule cells were host to the bioluminescent reaction, while the sticky beads that hang from the roof of the Glow-worm Tunnel were the points of light emission. No sunlight or any other light source was required to fuel this luminous display – light waves were *formed*, not reflected, from the energy released by a chemical reaction. That energy was light.

Unlike a light bulb, the chemical reaction involved in bioluminescence loses none of the energy input as heat, making this perhaps the most efficient energy-emission system known. It is even known as 'cold light', and that explains why bioluminescent fungi made good lanterns in a hay-filled barn – there was never a chance of sparking a fire. Yes, the fungi too were bioluminescent – they host similar chemical reactions within their cells, acquiring the oxygen component from air. But to what animal did those bioluminescent cells belong in the Glow-worm Tunnel?

The animal in question was the larva of a fungus gnat (a small fly). The larva is worm-like, hence the common name 'glow-worm' ('worm' is a generic name for any worm-shaped animal and should not carry information on classification – this 'worm' bears no relation to an earthworm, for instance). This particular worm is extremely small and black, making it invisible within the dark tunnel without a torch. Turn on a torch, however, and all is revealed.

Against the brown earth walls of the tunnel, the teardrops are visible as tiny transparent beads, draping from the roof like miniature chandeliers. Each bead is smaller than a drop of water, and twenty or so are evenly arranged along a fine silken thread like that of a spider's web. The thread is spun by the worm, which lives in a tube spun from the same silk on the roof of the cave. The worm intermittently secretes a clear, sticky fluid that drips slowly down the thread to form the beads, or teardrops. The fluid acts as a strong glue to flying insects such as flies, including the adults of its own species, and anything snared in the glue becomes the worm's prey. To lure its prey, however, it resorts to light. Bioluminescent light. The worm produces light that shines down the tube and threads and emerges only from the teardrops, as discovered by NanoCam.

Since my trip, the Glow-worm Tunnel has appeared in Australian tourist brochures. Connecting roads have lengthened and improved.

Hopefully the Rangers will monitor visitor numbers and restrict the use of torches to maintain this delicately balanced ecosystem.

Glow-worms exist even more spectacularly in Waitomo, New Zealand, and these have successfully borne the brunt of a century of controlled tourism. Dr F.W. Edwards, an early tourist and biologist who helped with our primary knowledge of the fungus gnat, painted a particularly lyrical picture of the Waitomo caves in 1924. As he stepped cautiously into a lower cavern, as part of a guided tour, he recorded: '. . . Then gradually we became aware that a vision was silently breaking on us . . . A radiance became manifest which absorbed the whole faculty of observation – the radiance of such a massed body of glow-worms as cannot be found anywhere else in the world, utterly incalculable as to numbers and merging their individual lights in a nirvana of pure sheen.' A wonderful depiction.

Bioluminescence is not confined to fungi and flies. Just to begin, the generic term 'glow-worm' was not even invented for the larva of the fungus gnat, but rather for an animal that makes a worm-shaped glow.

All over the globe, glowing worms can be found in open environments at night. They could not be more obvious – luminous spots cover either their whole bodies or parts of their bodies and shine out for tens of metres against dark backgrounds. Their lights are usually greenish yellow, and are turned on when disturbed, probably as a warning message that they are distasteful. Compared to the fungus gnat larvae, these glow-worms are large, the size of an average caterpillar, although there is some variation because they belong to many different species. They are the worm-shaped larvae of beetles. But within this group of beetles are other species even more famous still for their bioluminescence. These species become well known when they are adults and are the animals probably most synonymous with bioluminescence – the fire-flies.

Fire-flies

In terms of diversity and adaptability, beetles rule the world. Nearly one third of all known species on Earth are beetles. Three hundred thousand species of beetle are known, and probably the same number again or more have escaped our attention. Generally they possess two pairs of wings, although the forewings have become modified into hard, protective structures that must open before the hind, flying wings beneath can be deployed. Some species of beetle, on the other hand, have lost their aerodynamic wings and evolved a flightless existence. These include the long-legged darkling beetles of the Namib Desert on the south-west coast of Africa, which occupy the hottest terrestrial environment on Earth.

At the other temperature extreme, beetles can be found in ice caves in Canada at −83° Centigrade. These have evolved antifreeze chemicals, which are similar to the heat shock proteins found in hot desert beetles. And the beetles' size and shape portfolio is as impressive as their temperature triumphs, thanks largely to a suit of armour, known as the 'exoskeleton' (external skeleton), that is not only strong but also light-weight and crack-resistant. There are goliath beetles the size of a computer mouse, and rhinoceros beetles that can match the goliaths in length due to their elongated horns. Leaf beetles and Pie-dish beetles can compete with any animal for their bizarre forms, while diving beetles, although of run-of-the-mill beetle shape, have conquered fresh water. Then there are weevils, ground beetles, ladybirds and other groups that collectively feed on almost everything there is to eat on land. But of interest to this chapter is their remarkable chemical arsenal, where bombardier beetles can make explosions, and fire-flies can make . . . light.

Lasers are one case where man has advanced over nature. A laser beam is a light of a specific wavelength that travels in only one direction and is coherent (to some extent) – the rays are in phase. Animals lack the power needed to produce a laser beam, although the bombardier beetle does take a leap in power generation when it produces its chemically driven explosions (but a leap nowhere near far enough). Fire-flies, on the other hand, *can* emulate the effect of a human invention – the LED (light emitting diode), as found in coloured digital clocks. Both produce light that travels in all directions and is incoherent – the rays are not in phase. They are also easy on power. LEDs require only a low electrical input, while fire-fly bioluminescence involves a reaction similar to that of the fungus gnat, where the ATP in a cell can supply enough energy to drive the conservative, bioluminescent reaction – again, no energy is lost as heat. Indeed, this is one reason why white LEDs are poised to replace Edisonian bulbs, which require ten times more energy input because most of this is lost as heat (also, Edisonian bulbs have a tenth of the life of an LED, although they are currently a third of the price).

There are many species of fire-flies, or lightning bugs as they are sometimes known. They generate spots of light from different parts of their bodies. In some species the luminous pattern is a single spot on their body; in others there are pairs or rows of spots. The spots derive from bioluminescent organs known as 'photophores', accommodated within or near to the exoskeleton of the beetles. There is some variation in the design of their photophores on a microscopic scale, but all contain bioluminescent cells like those of the fungus gnat, and have the branching tubes of the trachea within or nearby bringing oxygen for their chemical reactions. An interesting addition, though, is a reflector.

A fire-fly's photophore has the form of a hemispherical bowl, occupying a cavity within the beetle's exoskeleton. The bioluminescent cells themselves, which occupy much of the organ, lie immediately behind the transparent outer surface or 'window'. And in turn, lying behind the bioluminescent cells of the photophores are reflector cells, lining the back of the photophore bowl.

Distributed *randomly* within the reflector cells' cytoplasm (the watery fluid filling the cells) are tiny granules of the base chemical guanine. The granules are just larger than the wavelength of light in size, and have an effect on all light rays. Light rays bump into, and reflect from, the granules. They strike the granules at different angles, and so bounce from them at different angles too. Some rays hit a granule head on, bouncing back on themselves, while others are just deflected slightly from an edge. The combined effect is that light rays of all colours are reflected equally over 360°. They are spread out into all directions, like the reflection from a pigment, and in white light the reflection will appear white from all angles. This optical effect is known as scattering, a form of structural colour because it involves physical structures (in this case granules), and is a common colour factory in nature that will re-emerge in this book. So when the bioluminescent cells emit light, some rays leave the photophore directly through its window, while others head towards the reflector cells. The granules reflect many of these rays back into the organ and again out through the window.

Behind the reflecting cells lies a layer of the black pigment melanin. The melanin soaks up those bioluminescent rays that make it through the reflecting cells, and so protects the beetle's tissues beneath from harmful intensities of light. So light leaves the photophore only through its window, and the bioluminescence that enters the environment is amplified.

Interesting that I compared a fire-fly's lights to an LED, because LEDs have been used by researchers at Tufts University in USA to fathom out *why* fire-flies produce bioluminescence.

The purpose of the fire-flies' lights

At the height of their exhibitionism, fire-flies light up sizeable areas of foliage like Christmas trees. In particular, there are mangrove trees in the State of Jahore, Malaysia, that turn on and off as if activated by a switch. To begin, a single flash from a single fire-fly triggers a tuning of the entire orchestra. Instantly the trees twinkle, randomly. Then order is restored and the light symphony begins.

In one area of a tree, the lights begin flaring in unison. Quickly the fire-flies in the remaining areas pick up the rhythm and the whole cluster of trees flashes on and off in synchrony. When they turn on, the lights are so numerous that it is not obvious whether one is observing bright lights against a dark background, or dark patches over a brightly lit background. But the interesting question is *why* are these beetles illuminating the Earth's surface at night? Well, we have cracked their code in a relative of the Malaysian fire-fly.

I subconsciously chose an orchestral metaphor to describe fire-flies because I first encountered them during a classical concert, in an open-air theatre near Washington, DC. The concert began in the early evening, and I sat with other visiting researchers from the Smithsonian

Institution in the cheap seats – a rug on the hill beyond the seating area. As the concert progressed, the sunlight faded, and soon the fire-flies emerged, first in the woodland surround and then above my head. They were spectacular, flashing yellowish lights to form collective patterns in the sky. The locals were used to them – they had collected them in jam-jars as children. But I could not even remember what symphonies had been played that evening, so distracting were the light displays. And now I associate fire-flies with classical music. Their lights, however, are not triggered by sound. Enter the LEDs.

LEDs have the effect of tiny light bulbs except that they don't possess a filament that can burn out. Instead they are made of a semi-conductor (a material that can conduct an electric current to various degrees) with an added metal compound. The precise proportions of the semiconductor and metal compound in the construction are so that an electric current can pass in one direction through the material, while causing electrons in atoms to jump to new energy levels, or orbitals, and instantly fall back down. In the same way as some pigments function, as the electrons fall down to their original (stable) orbital, they emit their energy as light of a specific wavelength. But unlike pigments they do not heat up, because all of the energy absorbed is re-emitted. This is a character shared with photophores, along with an enveloping surface that reflects the light produced outwards, into the environment.

The US researchers took yellow LEDs into a field inhabited by fire-flies. During previous nights they had filmed the native bioluminescence and noticed a pattern of flashes that seemed to recur. Now they could recreate that flash pattern using their controlled LEDs.

As the yellow LEDs emulated the fire-flies' precise pattern against the darkness of night, the fire-flies themselves approached. It became clear that the flash pattern served to attract members of its own species. The attracted individuals were captured and analysed, when it became clear they were all females. It emerged that males of this species produced the flashes and the females recognised and followed them. A good way to find and identify one's own species for mating purposes. In fact this experiment had confirmed previous work on fire-flies, although it did additionally demonstrate, by varying the effect of the

LEDs, that the females were not too particular over the precision of the lights, and would also follow a light that was larger, and longer lasting, than that of the male. It was discovered that the light signal indicated some quality of the males' 'nuptial gift' – the reproductive package of sperm and nutrient proteins. The longer the flash duration, the more robust the nuptial gift. Not surprisingly, female fire-flies were attracted to LEDs with an unnaturally long flash duration, stretching the boundaries of 'sexual selection'.

Earlier, pre-LED studies on fire-flies employed less-photophore-like light bulbs, but involved more intriguing species.

There are about 100 species of fire-fly in the eastern half of the USA. The males of one common species, the 'J-beetle' (*Photinus pyralis*), do not even wait for the dark – they begin their search for a mate at sunset, although in areas of deep shade. As the sun disappears completely, they move out into the open and emit a half-second flash of yellow light in mid-flight. During this light emission, they first dip in the air then pick up again, writing a luminous letter 'J' in the sky. As the flash finishes they continue their flight path, and six seconds later they write another J. They continue writing Js in space, usually next to vegetation, the possible resting place of a female J-beetle, while hoping for a reply.

The female J-beetle looks similar to the male, with equally well-developed wings and wing covers, although she has smaller eyes. But her main difference becomes evident at night – while the male lantern occupies the underside of two of its rear body segments, the female lights up only a crescent shape in the centre of one segment.

When the female notices a luminous, yellow 'J' in the sky, a timer is started in her brain. After two seconds, her brain sends impulses to the tracheal tubes in her photophore. The tubes flare, oxygen floods into the bioluminescent cells, the chemical reaction is fuelled and a flash of light is emitted.

The male detects the female's luminous response to his signal and traces the source of the light. Mating follows.

Other species of fire-fly in this region of the world have their own, equally distinctive flash patterns. For instance, the 'Dash-beetle' (*Photinus consimilis*) writes a pattern of three dashes (---) or nine dashes (---------), the 'Dot-beetle' (*Photinus ignitus*) produces a sequence of single spots (••••••), and the 'Zigzag-beetle' (*Photinus granulatus*) produces a pattern of zigzagged lines (ʌʌʌʌʌʌ). Each of these species repeats its patterns every five to six seconds, and it is always the males who perform the sky-writing.

The fire-flies listed so far all belonged to the genus *Photinus*. There is another genus, *Pyractomena*, that follows the same trend – one species, one flash pattern. This is convenient for biologists because the species of these genera appear very similar in body form, so the flash patterns are used as a means of identification. On a less selfish note it is useful for the fire-flies too – they can locate members of their own species at night even though there may be several different but similar-looking species inhabiting the same field. The bioluminescent patterns ensure that no mating mistakes are made. Well, one or two pitfalls do lie in store.

First of all, there is the chance that a fire-fly of the genus *Photinus* or *Pyractomena* may 'fall' during a light performance. In mid-air, half-way through a flashing sequence they may disappear – no more flashes and no more fire-fly in its projected flight path. With all that trouble of evolving a night-time existence, away from competitors, and a most efficient means of communication, this apparent failure in their system would appear unexpected. Conveniently, we can turn to an experiment to reveal all.

The bioluminescent display of a fire-fly can be imitated by dangling a small, flashing, yellow LED from a long cable suspended in the air like a washing line. The light should be free to slide down the cable, and it will imitate a signalling fire-fly rather well. But soon the light becomes a little dim. The reason for this is that it attracts and becomes covered in real fire-flies, but not those corresponding to this particular light display.

There is another genus of fire-flies in eastern USA – *Photuris*. Members of this group do not follow the rules. They are generally larger than other fire-flies, have longer legs, more agile bodies and will bite if handled, while flashing all the time. They use their biolumines-cence as headlights; for manoeuvring through grass, for egg-laying, and during take-off and landing. *Photuris* are also skilled in detecting the flash-patterns of other fire-flies. And they will bite the beetles that make them too.

It was the predatory species of *Photuris* that surrounded the artificial light in the moving LED experiment. When they spot a fire-fly's display from their base on the ground, they launch themselves like a light-seeking missile, homing in on the bioluminescence, whose conspicuousness is now beginning to backfire. The *Photuris* (predator) is then able to grab the naïve flasher in mid-air with its long legs and cause a stall, with the reward of a meal when both hit the ground. A simple strategy, in reaction to an obvious evolutionary selection pressure (one could almost predict the evolution of a predator that could make use of such a noticeable light), but that's not the end of it. The predatory fire-flies have cracked the code of the prey species, and can take part in male–female conversations of flashes as if they too were members of the prey species. The predators are able to speak the language that draws a prey species directly to its lair. As the male prey species homes in on the female flash pattern, his genes are passed on only to the stomach of a deceitful predator.

Fortunately the outcome of these conversations is mixed. Sometimes the male prey fire-fly *is* in luck, finding a real female of its species. And it's no bed of roses for the predatory fire-flies either. They also eat each other, and copy each other's distinct bioluminescent conversations in just the same way. Even mating is a dangerous business, in the style of praying mantids.

Cracking the fire-flies' flash code has not only led to the description of new species, but has also revealed a nice story of behavioural evolution. In the words of Charles Darwin: 'He who understands baboon would do more towards metaphysics than Locke' (Notebook M, 16 August 1838).

Humans take advantage of the fire-flies' lights

The idea that we could profit from natural bioluminescence followed a light-related human impediment – a glowing infection.

Two Australians were taken to the Gold Coast Hospital with spots on their legs that had developed into painful ulcers. Researchers at the hospital were called in when something unusual happened – the fluid from the skin lesions was plated on to Petri dishes, as normal, but as the room lights were turned off one night, the dishes were seen to glow in the dark. This was not normal, perhaps slightly worrying even.

It turned out that the patients had touched the bacteria *Photorhabdus*, the only bioluminescent bacteria living on land. This microbe hosts the luciferin–luciferase, bioluminescence reaction in its single cell. It also causes infections in humans, and its luminosity in this case made identification easy. After treatment with antibiotics the infections cleared, but there is an interesting lesson here that bioluminescence can be used as an eye-catching indicator of disease.

Currently, researchers aim to trace pathogens of crops using a luciferin reporter gene system. The approach is that the genes coding for the production of a luciferin and a luciferase should be added to the natural genes in the chromosomes of a crop plant such as wheat. The genes would then be activated, via a change in cell chemistry, only in the presence of a natural pathogen or pest, which would inadvertently switch on the entire bioluminescence reaction. This could represent the ultimate early-warning system of pests or disease for farmers – at night, the section of their crops under attack would glow. That part of the field could be isolated quickly, nipping the problem in the bud.

Although this idea has yet to be realised, the principle of genetically engineering an ordinary plant into a bioluminescent one has. In 1999, Professor Chia Tet Fatt from Singapore's National Institute of Education completed a nine-year project when he successfully transferred the fire-fly genes for luciferin and luciferase into orchid tissue. The transformed tissue was then propagated, and the resulting transgenic orchid – roots, stem, leaves and petals – produced a constant green light for up to five hours.

This feat was accomplished by first firmly attaching the genetic code for luciferin and luciferase to tiny beads. The beads were literally fired into the orchid tissue using a modified gun. They were able to break through the tough cell walls within the plant and deposit the genetic coding material for luciferin and luciferase into the nucleus of the cell, where it became incorporated into the orchid's genes. The outcome was that the orchids made a luciferin and a luciferase, and glowed in the dark accordingly.

The glowing orchids were selectively mated so that their offspring were also bioluminescent, and an industry was born. However, the bioluminescence is not particularly bright, and the effect would have been barely visible were it not for an astute choice of host orchid species.

Orchid breeders chose *Dendrobium* 'White Fairy Number 5', a species filled with a white pigment. Just like the reflector underlying the bioluminescent cells in the fire-fly photophores, the white pigment served to redirect bioluminescent light heading further into the plant back out, away from it, thereby increasing the intensity of the light observed. So the orchid appears green and mysterious at night after all. The perfect background for any outdoor soirée, were they not selling for $200,000 each.

Of course the production of a transgenic organism is an ethical issue, at the centre of current debate. For instance, while it may be possible to splice the luminescence genes into a plant, the glowing plant may be 'miserable'. Well, not such a tragedy for a plant, but the debate was further heated when the next model to be chosen was a rabbit. Fortunately the considerable expense required to make a bioluminescent animal cannot be offset by sales.

All the cases of bioluminescence considered so far have been on land, yet the Philippine blue, and the problem to be solved in this chapter, emerged from the sea. Moving from land into the shallow depths, a similar flashing repertoire to that of fire-flies can be observed, a scenario that also involves bioluminescent bacteria. However, although most of the bioluminescence encountered on land has been yellow or green, in the ocean it is usually . . . *blue*.

Flashlight fish

From the shore, alongside the shallow reefs fringing the Gulf of Eilat in the Red Sea, a diffuse blue glow can be seen covering an area of several metres. Diving beneath the surface in this region, into the centre of the glow, the cause of the light becomes obvious – flashlight fish (*Photoblepharon*).

The flashlight fish possess headlights. Their bioluminescent organs, or photophores, are positioned just below their eyes, and like the fire-fly photophores, they can flash. But their similarity to fire-fly photophores ends here.

The photophores of flashlight fish are in fact lit permanently. They manage to produce flashes thanks to black shutters that can completely cover the light source. When the shutters are quickly opened and closed, a flash of bioluminescent light enters the water. But why should such a complex system evolve to provide a flash? Surely the fire-fly system is more energy efficient, where their lights are turned on only when a flash is required?

The flashlight fish's lights are switched on permanently because they do not belong to the fish. Packed into their photophores are not fish cells, but bacteria – bioluminescent bacteria. The fish cultures the bacteria in a symbiotic relationship – the fish supplies the bacteria with nutrients and oxygen, and the bacteria provide the fish with a light source. So rather than controlling the bacteria's light production, an impossible task to perform with precision, it is simply covered when not required. A little over-complicated in theory, but it does permit a more disparate repertoire of flash patterns than that of the fire-flies, a potential fulfilled by the flashlight fish.

Flashlight fish produce three highly contrasting patterns of biolumi-nescence. First is a near continuous light with intermittent short, dark breaks, caused by infrequent closing of their shutters or 'blinking'. On average about three blinks are made per minute, each lasting just a quarter of a second. This light pattern is used to attract shrimp-like ani-mals living in mid-water that have good vision and are attracted towards light. Additionally the light from this pattern is used as a head-light, to see what lies ahead. The second pattern takes the form of 'one second on, one second off', flashing to a two-second rhythm. This pro-vides a means of communication for the flashlight fish, where the language may be aggressive, to warn off predators or defend territory, or it may be friendly, to initiate shoaling or mating. The third biolumi-nescent pattern involves the fish swimming slowly with its lights displayed, then suddenly the lights are covered and the fish darts in a different direction. The lights are displayed once more when the fish is in a different location, not predictable from its original, illuminated course. This 'blink-and-run' behaviour is performed over areas of reef offering little or no protection, and its purpose is likely to confuse or trick predators as to the fish's whereabouts.

Before learning of flashlight fish, I had considered a possible deceitful role of flashing lights with respect to the host's projected path. Returning home from work late one night, travelling within Sydney harbour on the outdoor seats of a ferry, I noticed an aeroplane passing overhead. The moonlight was particularly dim and I could see nothing of the aircraft itself, but I presumed it was there from the flashing lights in the sky. The lights attracted my attention in the first place, but then subconsciously I used them to guess the plane's projected path, so I knew where to look in the sky for the next flash.

Flashes were emitted in pairs. Those of a single pair were separated by just a split second and, importantly, each emerged from a different lamp – one at the front of the plane, one at the rear. One light flashed, then instantly the next, writing two dots in the sky. But observing these dots left me looking in the wrong part of the sky for the next. As a reflex action my head turned in the direction in which the dots appeared to be moving. The problem lay first in that my out-of-control brain had assumed that only *one* lamp had written the dots, and second in that they were written faster than the plane was moving. What actually happened was that the lamp at the front of the aircraft wrote a dot first, which was followed quickly by a dot from the rear lamp. But the rear lamp's dot had failed to catch up with that of the front lamp. So while the plane had moved from right to left, above my head, the dots had moved from left to right. Since I could not see the plane in the dark, my eyes moved in the direction of the flashing dots. The appearance was of a *single* lamp flashing as it moved from left to right. I had been tricked, and there was nothing I could do about it.

Now I look out for animals with bioluminescent lights at either end of their bodies that may recreate this illusion. Maybe, when in hot pursuit, some mid-water animals can turn to such a tactic to make predators think they have zigged when actually they have zagged. There are plenty of other bioluminescent fish that may fall into this category, although in these cases we know almost nothing of their behaviour and so cannot comment on the role of their lights. But at least the flashlight fish approach this illusion with their 'blink-and-run' behaviour. And then there is a squid (*Chiroteuthis imperator*) that dangles a long, thin tentacle with photophores that flash in an orderly row . . . Researchers must hope for more submarine time.

Placing theory aside and returning to the problem central to this chapter, could flashlight fish lie behind the blue light that lit up the Philippine bay? Well, probably not. First, the light field produced from a shoal of flashlight fish appears to flicker, since the individuals within the shoal all flash their lights rather than leave them on – the Philippine glow was uniform and continuous. Second, the Philippine lights stretched to the very boundaries of the bay, ending on the beach itself. The flashlight fish, on the other hand, occupy reef areas where they find abundant food, but would not venture so close to the shore to avoid beaching. They would have left a margin of non-luminous sea at the edges of the bay. But there are many more cases of bioluminescent marine animals to examine. Most of these belong to the deep sea, which I will save for the Red chapter, along with a rather unforeseen peculiarity. After all, the blue glow in question in this chapter occurred in a Philippine *bay*, so it is the *shallow* seas that should be trawled for evidence.

Seed-shrimps – fire-flies of the sea, and more

Crustaceans are often termed the insects of the sea. This analogy is echoed in their general small size, variation in body form, their diverse range of habitats and lifestyles, and their domination in terms of species numbers and biomass of, in this case, the ocean. As members of the great 'arthropod' phylum, which contains more than three-quarters of all animal species, they are protected by an exoskeleton and *are* related to insects. Their most famous affiliates include shrimps, crabs and lobsters.

A luminescent crab at night is an eerie sight. I spotted one on a beach on Australia's Great Barrier Reef – a crab-shaped blue light, roving an otherwise dark beach. I investigated. The crab was transparent and actually not bioluminescent at all, but its dinner was. The crab had eaten another crustacean, a seed-shrimp, which had produced bioluminescence within the crab's stomach. Since the crab's body was transparent (all except for its eyes), hence its common name 'ghost crab', the seed-shrimp's bioluminescence shone through its exoskeleton. This was ironic because I had recently found a minute but strikingly colourful opal that turned out to be stuck in the stomach of a transparent seed-shrimp – the seed-shrimp had eaten the opal.

Bioluminescent seed-shrimps are tomato-seed-sized animals with a shrimp-like body that is totally enclosed in a two-part shell. Their claim to fame was their role in map-reading by soldiers during the Second World War, although unfortunately they were killed before being put to work. A single seed-shrimp's bioluminescence has a very low intensity, but during jungle warfare that was ideal. It was just enough to illuminate small sections of a map, but not enough to make the soldiers stand out in the dark. So seed-shrimps inhabiting coastal regions were fished and sun-dried – when dead they retain their luminous chemicals, which mix and react when the body is crushed.

Today male 'notched' (cypridinid) seed-shrimps are more famous for their fire-fly-like dances on Caribbean reefs, where different species produce different patterns of bioluminescent flashes in the water column. The females are lured out of the sand by these dances and attracted towards the particular dance of a potential mate. But the evolutionary road to these flash patterns began with single flashes of bioluminescence and in larger amounts.

An innovative seed-shrimp biologist, Katsumi Abe, once passed by me an idea that he had pieced together through many careful observations. This idea held that the bioluminescent chemicals of notched seed-shrimps, which are squirted from glands in their upper lip, evolved from digestive enzymes, secreted as a mucus and used to partially break down food before it enters the digestive system. A nice hypothesis, of which there are few for the origin of bioluminescence in any animal. That a digestive mucus surrounds the bioluminescence of these animals to ensure a condensed, ball-like flash supports Katsumi's hypothesis.

Sadly and tragically, Katsumi died recently, while returning home from his lab late one night, but his students continue building on his scientific foundations.

The luciferin and luciferase of notched seed-shrimps are produced in different glands, and meet for the first time in the open water, where they react with oxygen. This bioluminescence is external to the animal itself, providing an option to produce a cloud of bright light in the water while the seed-shrimp makes a quick exit. Many notched seed-shrimps are known to practise this behaviour, causing a predatory fish to become momentarily dazed – the cone cells in its retina fire so frequently that somewhere along the visual processing line the image is saturated and the fish is temporarily deprived of its sight.

Evidence that this defence tactic works can be found in the collection method for flashlight fish. At night the fish can be seen easily from twenty metres away, and appear undisturbed at distances of up to one metre. But move any closer and they will take evasive action (their blink-and-run behaviour), so they can be neither grabbed by hand nor caught in a hand net. However, shining a bright underwater torch at them elicits temporary blindness. They can no longer see or, consequently, behave as normal, and in this state they *can* be captured by hand.

The only glitch I foresee in this 'smoke-screen' strategy is that by dazzling a nearby fish, the light produced will also *attract* the attention of a more distant predator, beyond the sight-deprivation zone. Then again, if the seed-shrimp can flee from its light quick enough, then by the time a distant fish arrives it could be long gone. Bright lights can certainly override the other stimuli, including sound and chemicals, but they can prevail over reason too, to elicit an irrational response in the receiver animal. The British military once discovered this when their visual sense overcame their map-reading skills.

In the dark morning hours of 10 August 1813, a fleet of British barges prepared to launch an attack on St Michaels, a town and a fort on the harbour side of the Chesapeake Bay, near Washington, DC. The residents of St Michaels, forewarned, extinguished all their house lights and moved to a point a little further along the bay, where they hoisted lanterns to the masts of ships and the tops of trees. The British saw red and reacted. Rather than sticking to their trusted maps, they aimed their cannons towards the lights, which in their minds represented the houses of St Michaels. The result is that they overshot the town (with the exception of one house, where a cannonball reportedly pierced the roof and rolled down the staircase). Proof that a bright light can not only distract but also trigger uncontrolled behaviour. A useful weapon, then, for prey.

Some seed-shrimps take the art of smoke-screens to an extreme. A different group, the 'notch-less' (halocyprid) seed-shrimps, have evolved bioluminescence independently. The glands producing their luciferin and luciferase are housed in their shells rather than their lips. And they have also evolved a very different function for their biolumi-nescence, which appears as a very different light effect.

The notch-less seed-shrimps congregate in their thousands at the surface of the ocean at night. During the day they evade their main predators by sinking over 100 metres below, but at night, when they rise, they are vulnerable to predation. And their defence strategy is quite surprising. Rather than keeping a low profile, they go to the other extreme and make themselves as conspicuous as possible – in fact the more conspicuous the better. They huddle together into a concen-trated area and glow for all they are worth. They all release their bioluminescent chemicals into the same patch of sea water, which accordingly glows blue. Against the black background, this blue light is intense.

As expected, the fish predators of the seed-shrimps notice the blue glow from afar, but unpredictably they ignore it. They do not venture anywhere near the beacon, where they could be sure to find a feast. The reason for this is as much to do with size as it is with vision. The fish predators of the tiny seed-shrimps are themselves small and not top of their food chain (in its simplest form). They have their own predators; larger fish and squid for instance. And all the predators involved in this chain use their eyes to find prey.

What the notch-less seed-shrimps have evolved is a 'burglar alarm' strategy. Their concentrated area of bioluminescence has no gaps – it is a continuous light field. So anything in front of this light will appear as a silhouette, and a very obvious silhouette at that. The shape of a small fish 'cut out of' the bright, blue background would be projected many tens of metres through the ocean, and into the eyes of its predators. *Now* it is not so surprising that the small fish stay well away. By making themselves conspicuous, and without any defences (this is not a warning colour), the notch-less seed-shrimps have evolved an excellent and unique anti-predator strategy. And their lights may additionally add to a mating strategy, bringing related individuals together.

Returning once more to the problem of this chapter, could notch-less seed-shrimps be the cause of the Philippine blue? This is why the Assistant Director of the Bureau of Fisheries and Aquatic Resources, Republic of the Philippines, wrote to *me* – I work on both seed-shrimps and light. He was hoping for a 'Yes' to this question and to close the file on his very public enigma. I answered, 'No.'

The only records of notch-less seed-shrimp lights were from beyond the shore, in regions of deeper water. This makes sense because the animals must migrate downwards during daylight. Whether they also produce their clouds of light while in the deep water, which is quite dark during the daytime also, is unknown – again we need more submarine time. But we can be sure that the blue light of the Philippine bay had a different origin. And I knew what.

The solution – 'sea-sparkles'

I solved the problem central to this chapter while on another excursion with Peter Serov in 1994.

Again we set out from the Australian Museum, but this time in a Toyota Landcruiser and at night. Shane Ahyong, another friend and colleague in the marine invertebrates division, joined us. We were heading to Botany Bay, the place where Captain James Cook first set foot on Australian soil over 200 years ago, just south of Sydney city centre. This was to be my first night dive.

We parked at the edge of the vast bay and began unloading our dive gear when I noticed a sign. It was obvious – the white, wooden board was directly lit under a street light. And it was to the point, with just a single word, written in large, red letters, and a simple picture of an animal, painted in blue. The letters read 'DANGER'; the picture was of a shark.

Sharks come in all shapes and sizes, and most do not attack humans. But the shark in this picture looked menacing, with a large, gaping mouth. Although the rest of the animal was drawn in minimalist, almost caricature form, the artist had taken the trouble to add teeth to the mouth. This was not what I wanted to see just before my first night dive. I looked out towards the water and it suddenly seemed different from my impressions from all previous occasions. The inviting, sunlit, crystal blue waters had transformed into a far more forbidding environment. I must confess that I am slightly more conscious of the threat of sharks than most divers, and my friends knew it. So the stories, in the manner of *Jaws*, began.

I became happier to enter the water when I saw a more calming object – a shark net. A section of the water, marked by floats at the surface, had been isolated by iron bars; effectively a titanic cage. I squeezed into my dive gear, but still appeared apprehensive, as evident from the photo taken. I was decidedly white, while Peter and Shane were smiling, with the manner of smile I reserve for 'the evil' Moriarty.

We snapped our light sticks, signalling that we were ready to go. Light sticks are used by night divers, but are also found as necklaces at night-time fairgrounds. They consist of a thick plastic tube with a smaller, thin-walled glass tube inside. Both tubes are filled with chemicals; different chemicals, similar to luciferin and luciferase. When the outer, plastic tube is bent, the glass tube within breaks, and the two chemicals mix and react. The reaction, known as 'chemiluminescence' (where it has no biological origin), gives off light as a by-product. Again, no heat is emitted.

With light sticks tied to our buoyancy jackets, we broke the still, moon-reflecting surface of the water and looked beneath.

The first thing to notice was that the light sticks really worked. I could see exactly where my dive buddies were in the otherwise pitch-black environment. My second recollection, after the light-stick novelty wore off, was of the eeriness. This felt like a place I should not be, as I soon realised other animals should. I spot-lit the sea floor with my

Figure 4.1   Shane Ahyong (back), Peter Serov (right) and me just prior to entering the water on a night dive. Sydney, 1994.

dive torch and saw long 'ribbon' worms moving in all directions, and an octopus chasing a large crab. In fact a whole range of invertebrates could be seen, from a whole range of major animal groups or 'phyla'; a very different sight to the fish-dominated daytime. After the wonder of a new world (to me anyway), I was brought back to reality with a return to the eerieness. I shone my torch ahead of me to reveal the safety bars separating this section of the sea, and noticed immediately that several bars were missing. Several *next to each other*, leaving a gaping hole the width of an adult tiger shark (my units of measurement give some indication of how I was thinking).

I got on with the dive and saw exactly what I was looking for – bioluminescence. A chain of small, free-floating, box-shaped animals known as 'salps' (actually lying just within the chordate phylum to which we belong) appeared, like glowing pomegranate seeds drifting together in a row. Suddenly I was happy again, and followed the blue lights around the supposed enclosure.

Eventually my eyes fully adjusted to the darkness, and I saw more. More bioluminescence and, finally, the answer to the Philippine blue mystery.

As I resurfaced I noticed that my light stick was not the only lumi-nescence indicating my presence in the water. I stretched my arms out in front of me, and their outlines shone. My whole body was silhouet-ted against a blue glow, and my dive buddies were now ghostly figures suspended in the water too. Unlike the notch-less seed-shrimp biolu-minescence, there was no large patch of light. The fact that it took some time to even notice this light meant that it was dim, but certainly it was a light triggered by our presence. I took a sample of the glowing water, rushed back to my lab at the Australian Museum, and examined it under the microscope.

The light, which could clearly illuminate any object, no matter how large, had emerged from a single-celled creature. A drop of this sea water on a microscope slide revealed tens of spherical cells, each a 'dinoflagellate' known as *Noctiluca* or, commonly, a 'sea-sparkle' (cur-rently, we do not know the precise classification of dinoflagellates – they belong to part of the evolutionary tree that is in a state of aca-demic transition). The sea-sparkle takes the form of a microscopic bag, which floats right under the surface of the ocean. It contains a luciferin and a luciferase that mix together, and produce bioluminescence, when the organism is agitated. The movement of my arms had provided such a stimulus.

Although the concentration of sea-sparkles in Botany Bay had been low, they sometimes occur in great numbers, around a million individ-uals per litre of water. And that's precisely what had happened the other side of the equator in the Philippine bay. They had been drawn in towards the land in a current, and formed an impromptu aggregation, with the beach preventing their further movement. They had been channelled into a corner, from where they were destined to wait for the tide to carry them out again.

Sea-sparkles exist at the very surface of the water, and the motion of the ripples on the water's surface is enough to agitate their cell, result-ing in a bioluminescent emission. And in that Philippine bay, a million sea-sparkle cells existed per litre of water. The consequence of that was a mass light bomb, or the '*Philippine blue*'. We have our answer.

The reason the blue comes and goes in the Philippines depends on the locality of these animals in the ocean and on the tides. They are carried by the tides, so if they drift close enough to a current leading into a bay, then that is where they will go. They will leave the bay when the tide is reversed.

Sea-sparkles also become cornered in other bays of the globe, such as Puerto Rico's famed 'Phosphorescent Bay', where they leave their greatest impression on the Earth's surface. But they make themselves known, generally in rough or choppy surface waters, all over the world.

Two thousand and three hundred years ago, Aristotle wrote, '... Some things, though they are not in their nature fire nor any species of fire, yet seem to produce light.' He was referring to sea-sparkles, which are seen by all sailors as they illuminate the crests of waves and the oars of rowing boats, becoming particularly visible when the moon is least bright. According to the legend, the last submarine to be sunk during the Second World War was detected at night, while at the surface, by virtue of its luminescent wake. And sea-sparkles continue to provide military headaches today. No matter how stealthy a ship is designed to be, if it sails through a patch of sea-sparkles it can be detected from beyond the Earth's atmosphere. An impression that embodies the effect of bioluminescence and the need for continued visual adaptation into the night, and an interesting note on which to end bioluminescence, at least for land and shallow seas.

A tonic for Darwin

The themes of this chapter were bioluminescence *and* darkness. The two are complementary. But why?

Mushrooms were glowing at night possibly to attract insects, as were glow-worms in caves. Fire-flies produced glowing flashes to attract other fire-flies, and flashlight fish spoke to each other in a bioluminescent language, but again, only at night. The fact is that during daytime bioluminescence does not appear all that bright. It is never outstanding amid the sunlight that floods across fields or through the surface waters of the ocean. At night, on the other hand, the general light in the environment is considerably lower – moonlight is around

1,000 times less intense than sunlight. Under *these* conditions, bioluminescence comes into its own. Pigments and structural colours fail to work effectively with little light to fuel them, but bioluminescence requires no light source. And at night it is over 1,000 times brighter than the background moonlight. So the effectiveness of bioluminescence is all down to intensity differences – its contrast with the light levels in the environment.

The fire-flies and flashlight fish were adapted to the light levels of bioluminescence rather than moonlight. When their eyes detected a green or blue flash, their brain processed this sensory information and interpreted the signal. The flying insects in caves, on the other hand, were adapted to extremely low background light levels. When *they* saw a blue, bioluminescent glow it flooded their visual systems and the stimulus overcame their normal functioning. The insects flew towards the light, as if in a trance. This was not in the insects' best intentions, as soon they would become main course for a glow-worm. Flying insects are attracted to light bulbs at night in the same way. Indeed, even the flashlight fish, adapted to their own, relatively bright bioluminescence, could not deal with a torch at close range, representing a brighter light still.

Similar was the case of the seed-shrimps, at least most species of seed-shrimps. Their bioluminescence was employed to startle predatory fish at close quarters – an uncomfortable position for the tiny seed-shrimps. Their bioluminescent clouds were just too bright and caused the fish's visual systems to become momentarily dysfunctional.

Now it can be understood that bioluminescence demonstrates the third lesson in 'eye' imperfection – that the visual system can be adapted to a *narrow* range of intensities only. The minimal levels of light in the animal's environment usually dictate this range because these tend to be the most frequent, background levels. General day-to-day seeing, therefore, involves the detection of light of these levels, and the eye evolves to be adapted to them. That the eye can detect only a narrow range of intensities, however, means that a light much brighter than the background illumination will bring about problems. It will lie above the threshold value for the visual system, causing vision to break down.

The visual system contains an element comparable to fuse-wire in an electrical circuit, but with a minimum as well as maximum level at which it can function. A thick fuse-wire will conduct a wide range of currents, but a thin fuse-wire will break or 'blow' when the current becomes too large. In comparison, the visual system will become 'saturated' if its signals surpass a certain level. Not only will it fail to distinguish between intensities of light detected above this level, but it will fail to do anything useful with the visual information at all. The brain becomes akin to a video camera monitor when the brightness levels are set too high – it blanks out with a flood of light. The entire visual system then takes a moment – maybe seconds – to reset and to work again. Seconds, in nature, can be critical. In the case of the seed-shrimps, the seconds taken for a fish's eye to function once more following a bioluminescent 'blast' allow a seed-shrimp's escape. Unlike this predatory fish, the fire-flies were adapted to the top end of the light intensity range. They detected their own bioluminescence, but would have been oblivious to signals involving the reflection of moonlight.

A visual system that could manage light of a wide range of intensities, like a thick fuse-wire, would be the ideal. That would be the *perfect* visual system. But such a visual system is never found in nature. The 'eyes' of animals have the 'thin fuse-wire' scenario, and a bright light will cause their fuse to blow and their visual system to be found *imperfect*.

Some animals, like ourselves, can adjust to different light levels. However, the switch from dim to bright light sensitivity cannot be made immediately. We do not have a 'thick fuse-wire' but must select the category of light levels in which we would like our narrow range of sensitivities to function. If we encounter an instant bright light while in low-light detection mode, we will be dazzled. Consider an oncoming car at night with headlights in full beam – that same beam would be far less dazzling if viewed on a bright, sunny day.

Also interesting were the ghostly impressions that appeared in this chapter in the eyes of Philippine people – the fungi in the graveyard and the sea-sparkles in the bay. These were cases where the visual processing system had effectively malfunctioned by incorrectly identifying the objects in sight. I will say no more about this here because more impressive cases of 'visual trickery' in the brain will emerge in later chapters.

Certainly in this chapter, the eye was revealed to be *not perfect after all*. On account of the eye being too *perfect* to result from evolution, once more *Darwin had no cause to worry*.

The dark will be left behind in the next chapter and another curious case of adaptation to both sunlight and eyes will be put under the microscope. But this adaptation was, 200 years ago, believed to be maladaptation, unless the little known Australian trees at that time were really clad in blue leaves. The characteristics of pigments and structural colours will become important, and those foolproof wires in the visual system may appear to become crossed for the first time.

green

The problem:
Why does the Australian 'blue frog' seek camouflage on green
leaves?

The findings of the First Fleet

In 1770, Lieutenant James Cook's *Endeavour* anchored in Botany Bay,
just south of Sydney. Having safely completed his primary scientific
assignment, to observe a transit of the planet Venus across the face of
the sun from Tahiti in the South Pacific, Cook's new mission was to
find the undiscovered southern land – Terra Australis Incognita. Many
Renaissance geographers believed this land to exist to counterbalance
the northern continents, and earlier European expeditions had hinted
of a great landmass between the Indian and Pacific Oceans. But the
coastline was completely unknown, with the eastern side known only
as 'a place where there be dragons'.

After following the southern coast of Australia, Cook was desperate
to re-supply the ship before returning to England. Botany Bay was the
chosen dockland for its exceptional size, and Australian natural history
began.

The ship's scientific officer was Joseph Banks, a qualified scientist
and one of the ablest young botanists of the day. Dr Daniel Solander, a
gifted pupil of Swedish naturalist Carolus Linnaeus, had joined Banks's
scientific party, who made a swift, biological dent in 'the bush'. Banks's

team became famous for the botanic specimens they drew with accuracy, collected and brought back to London. Leaves and flowers were pressed and dried so that they still survive today in the Herbarium collection of Kew Gardens – we can vouch that these drawings really were remarkably precise. British naturalists became instantly educated in the weird and wonderful flora of Australia that so astounded and delighted Banks, and the news soon spread around Europe. The terrestrial fauna of Australia, on the other hand, were less obliging to Banks and his colleagues – they tended to move.

Records suggest that Banks's party did notice the insects, in particular the 'troublesome' giant ants that draw blood when they bite, and the cicadas, which are unfamiliar to most people today during their first encounter. Cicadas can appear more like small birds than insects. Yet attention was reserved mainly for the birds and mammals that, nonetheless, were alien too.

Today it is only the *original* mammal drawings that seem alien, estimated from glimpses of wallabies as they turned rocky corners, and platypuses as they surfaced to breathe. In the artists' defence, they had seen nothing before to scientifically compare with these animals. I often wonder how people before zoos and (accessible) books must have interpreted some of the more enigmatic or giant creatures for the first time – the Egyptians as the Nile crocodile first surfaced at the Temple of Luxor, or the Ancient Indonesians as Komodo dragons investigated their settlements. In the Australian case, we are given a hint of such thoughts – we have the drawings of the first European tourists.

Banks's scientific reports and general recommendations for the European settlement of Australia started a very large ball rolling. Further expeditions from Britain followed promptly. The famous First Fleet arrived at the site of Sydney in 1788. These eleven ships were filled not only with convicts and officers but also with more scientists, including an astronomer and John White, a surgeon with a strong interest in natural history. Australian botany and zoology was about to become serious, although not without some initial mistakes.

Some of the early technical illustrations of kangaroos are barely more scientific than their interpretations in Aboriginal dreamtime paintings, which fully exploit their artistic licence. Kangaroos and wallabies emerged as large hares, standing in their familiar upright position within a British landscape, although with rat-like rather than bushy tails. Koalas appear like squirrels without their tails.

My comparisons here are appropriate. A Rosetta Stone can be found in the oil paintings of the first British artists of Australia, who, versed in the faithful reproductions of Constable's countryside, were capable of replicating sights with such realism that only the invention of the camera would wipe them out. Yet their Australian bush scenes were far from realistic. Eucalypt trees are characterised by tall, fairly straight trunks with rather unswerving branches. In these oil paintings their trunks and branches appear curvy and meandering like those of the British oak.

The artists were under the influence of their native imagery, purposely in the case of the oil painters (to remind the homesick emigrants of Britain) and subconsciously for the scientists. The scientists had fallen victim to a fault in the visual system, which came to the fore in the previous chapter. The images of Australia formed on their retinas were absent from the dictionary of their visual cortex, but rather than becoming new records, they were shoehorned into their closest entries. In this way something unfamiliar can easily become something familiar. The botanists, in contrast, had no such problems, since they were even able to trace their leafy subjects. They also had plenty of time to mix up the correct paint colours – the correct shades of green.

John White paid a little more attention to Australian animals than had Banks's party, and did make collections of the fauna, including a frog. A rather placid and completely harmless frog. To White, frogs were amphibians that lived in and around ponds, or at the very least on the ground. The frog he collected, on the other hand, had been prised from a tree. Caught in the excitement of a biologist's paradise, White's collection trips must have verged on the whirlwind, grab-what-you-can types. Rather than wasting time on detailed notes, the animals accumulated by White were hastily preserved, with a label recording only their locality and habitat, and casually shelved in his office. Haste that would, eventually, result in an embarrassing scientific error.

It was clear that, unlike the hard-shelled insects, the frog would quickly dry out and shrivel up. The art of preserving animals with the consistency of a frog was in its infancy, but White assumed from his medical training that alcohol would be key. Unfortunately alcohol for *this* purpose was not among the scientific rations of the First Fleet, but an interesting corner was cut. 'Alcohol is alcohol,' thought White. Along with its paper label, providing the New South Wales address down to the tree on which it was caught, the frog, no longer alive, was duly squeezed into a bottle of rum with enough of the alcohol remaining to cover it.

The rum survived the demands of the new colony, or specifically the temptation to dissolve one's problems in alcohol, and so the frog specimen survived too. It would seem that the frog's now grotesquely unnatural face peering out of the bottle was enough to deter even the desperate. Besides, the rum was beginning to cloud, and the aroma must have provided a final defence against anyone game enough to remove the cork.

A year or two later, John White returned to his rum bottle, specifically to apply the logic of Linnaeus. Carl von Linné (or Carolus Linnaeus) was the father of biological nomenclature, or 'taxonomy', through assigning 'scientific names' to his and his students' plant collection (including that of Daniel Solander). Each name came in two parts, both derived from the Latin language. The last part was known as the species name, the first part the genus name. All plants with similar characters were placed into the same genus, while only those 'identical' (within limits) were designated the same species name.

The choice of Latin names was, and remains, the privilege of the person describing the organism for the first time. During the process of nomenclature a particularly characteristic trait may be endorsed – something special about its environment, or perhaps the colour of the organism; on other occasions the collector or a peer is honoured. The English (or other) language becomes converted to Latin form and the scientific name is born.

Back in his office, John White blew the dust from his rum bottle, cracked the glass and lifted the Australian frog from the alcohol. Although all frogs were given the genus name *Rana* meaning, believe it or not, 'frog', the shape of this specimen differed so markedly from the frogs in European collections that it required a new species name. And

The pygmy seahorse camouflaged against coral. The following pages show different colour factories in animals at work. (See captions on page xi).

ultraviolet

violet

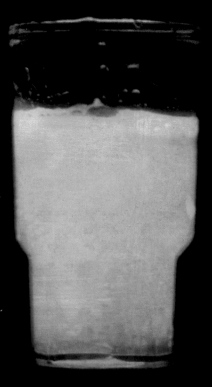

blue

green

yellow

orange

red

there was really little option but to add the Latin translation for a strikingly obvious and unique character that any fool could see – the blueness.

'Blue frog, speckled beneath with greyish; the feet divided into four toes; the hind feet webbed. Size of the common frog.' This was all White had to write about his amphibian. But European frogs were not known to be blue. Browns, greens, yellows, even patches of red, but never blue. So *caerulea* became the species name, from the Latin *caeruleus*, meaning the dull blue that seemed most representative (sometimes even known as 'cerulean blue' today). As this nomenclature entered the scientific literature with the 1790 publication of White's work *Journal of a Voyage to New South Wales*, the Australian frog became officially *Rana caerulea*, the 'blue frog'.

At around 1880, White's frog was transferred to the newly established genus *Hyla*, and finally, following further biological dividing, to *Litoria* in 1971. Today, its scientific name is *Litoria caerulea*.

A niggling doubt

The label drowned in rum insisted that the 'blue frog' was found sitting on a leaf within a forest canopy. Examination of its adhesive-pad-like feet further suggested that this frog was at home in the treetops. But Banks's botanical illustrations taught that the leaves on which the 'blue frog' sat were green, which should make the frog an easy target for predation. Clearly something was odd.

Even at this time, camouflage was well understood. It is after all rather self-explanatory, surely lying at the foundations of animal behaviour study. There can be no doubt why a stick insect is brown or a leaf insect green, even before their common names were added. So why should the blue frog – a defenceless chunk of protein – choose to sit among green leaves?

Generally, when viewing a field known to be packed with animals, from insects and spiders to rodents and birds, almost none of that fauna is observed. The reason for this is clear. Retinas exist in and around that field, and accordingly everything exposed in the field will leave impressions on them; a thought made more interesting when the retinas of *predators* are considered. So to avoid becoming easy prey, any animal in that field must adapt the impression it leaves on a retina. There are, fortunately, options for visual adaptation, which evolution has exploited.

Camouflage and artistic realism – the rules

In the statement above I claimed that almost none of the field's fauna could be observed. Well, that's not strictly true. Certainly some animals have adapted their behaviour to be hidden from sight. Wrens tuck themselves into dense foliage, woodlice (or slaters) hide under stones and fallen branches, and badgers venture out of their sets only at night, so all are literally out of sight. There are, on the other hand, more daring creatures that have evolved to shun protective barriers and trust in their camouflaged colours, shapes and behaviours, often employed in combination. So they *are* observed, in that they leave their images on retinas; they just match their background so well that they do not stand out as their true selves. I should have rather claimed that almost none of the field's fauna could be *identified*.

In the following several pages, I will make comparisons of nature with *art* in an attempt to highlight the complexity of camouflage. It is interesting how both have often converged on the same solution for a shared problem, but art may help to make these problems sound familiar because here *we* are the target species.

The evolution of camouflage (or mimicry for that matter) faces challenges similar to those of the artist wishing to retain at least an element of realism. In the case of camouflage, an animal evolves with the appearance of another animal or plant. The artist, with a goal in sight, strives to make one thing look like something else; the canvas must become a window through which, for instance, a real countryside scene can be viewed. The big, big problem facing both artists and nature is the visual system.

There are three visual elements to both the artist's and nature's challenges. The first is the countryside itself. This is nature, and the only honest element. Trees packed with green and brown pigments sway in the wind. Male pheasants emit green flashes as their structurally coloured heads stoop to feed. And red pigmented poppies provide a focal point within the bland grasses and so attract pollinating insects.

The second element is the visual system – the troublemaker. 'Trouble' because it does not faithfully record what lies ahead. As we have learnt in this chapter alone, early scientists drew kangaroos in the style of hares – a little dramatic, but all the same a case of the manipulative visual system at work. This case concerned a flaw in the brain, the interpretative part of the visual system. The brain's arsenal of decoding methods includes association. Many *characters* are associated with any single object, characters including colour, movement and surface texture. An orange must have the faintly dimpled texture of an orange. A leaf must sway in windy conditions but be motionless in still weather.

For the brain to identify an object, that object must possess all the characters associated with its definition in the brain's dictionary, like a set of numbers to open a combination lock. Problems do occur when the brain sees something for the first time, and there is no entry in its dictionary. In some of these cases, the brain can be supplied with most of the numbers for another, close entry but adds the final number itself, in order to open a lock and force an identification. In this way the green lights in the Philippine graveyards and blue lights underwater were interpreted as ghosts, and kangaroos became giant hares. These unauthorised goings-on in the brain can be demonstrated using Kanizsa's triangle, where three sectored-discs and some lines are set against an unvarying, evenly bright background. But we see something that is not in the figure – a bright triangle, seemingly lying over the discs and lines. The lines and the sectors in the discs offer enough characters in the brain's combination for 'triangle' for the brain to be satisfied that a triangle lies ahead. It adds the remaining characters itself. A pattern of three dots which, when joined up, form a triangle, supplies insufficient characters for the brain to find any entry in its visual dictionary. No triangle is perceived from just three dots.

I'm afraid we must accept that occasionally the brain becomes a loose cannon because this is inevitable with so many demands placed on it. Yet it still manages to astound those prepared to scrutinise it. The

Figure 5.1   Left, Kanizsa's triangle, containing many, but not all, characters required to identify a white 'triangle'. Right, pattern with too few characters of a triangle.

brain as a whole is greater than the sum of its parts, thanks to the little extra jobs it performs on its own accord, like adding pieces to incomplete puzzles. Computers, in comparison, adhere strictly to their instructions, and consequently their 'whole' is equal to the sum of their parts (to a degree). The difference between the brain and a computer is true intelligence, so the occasional fault in the brain is a small price to pay.

With a consideration of intelligence, it is now becoming clear how the visual system sets the rules for the characteristics of either a painting (resulting from the artist's hand), or a *camouflaged* animal (resulting from evolution). Both strive to appear like something else, and it is how that 'something else' – whether a bowl of fruit or a leaf – appears in the mind that sets the rules. In the case of the realist artist, the actual bowl of fruit *and* the paint on canvas must appear identical in the *human* brain. The camouflaged leaf insect, on the other hand, must appear equivalent to a leaf in the *mind of its predators*. In other words, the leaf insect need not reproduce the exact physical properties of a leaf because its predator's brain does not make use of all of these properties anyway, and maybe introduces additional effects.

So it is the image in the predator's brain that becomes both the model and the selection pressures for evolution – the leaf itself never enters the evolutionary equation. From this, artists can learn that trees painted on their canvases are to appear as trees in the eye of the observer *only*. They may be granted short cuts to achieving reality by understanding how the visual system works and perceives objects – the artist may cheat. Indeed, the Impressionist artists *did* cheat.

Background effects and camouflage shape

Another character used by the brain to identify an object is its background. Certain objects are associated with certain backgrounds, and if the background appears odd, then the combination lock remains unopened. The brain registers something unreal. This is the principle behind Surrealist art. Although each character is painted with accuracy, the mismatch of background association causes alarm bells to ring. Things that are not quite right in the background, such as the absence of shadows in bright sunlight or leaves beneath shedding trees, have the same effect.

Hanging in the National Gallery, London, is a pre-Surrealist-movement (1762) painting by British artist George Stubbs, called *Whistlejacket*. Immediately the painting grabs everyone's attention, not only because of its sheer size but also because there is something so obviously wrong with the subject that the painting as a whole makes one uncomfortable. A brown, rearing horse dominates the canvas at near life size. It is painted in fine detail and with remarkable accuracy – it looks so real (Stubbs was also a natural scientist, whose study of bones and muscles informed his canvases). But at the same time the whole painting conjures abnormal, even eerie feelings. The deliberate problem lies in the background – it is uniformly pale brown. We never view a horse against such a background; it is not one of the characters in our combination lock for 'horse'. To make matters worse (or better for this artist) the horse leaves no shadow. This is an excellent example of how background plays a necessary role in association and as a result identification. In nature the same trap is set for the decorator crab, which has evolved a behavioural response.

The decorator crab (*Libinia dubia*) of the eastern USA benefits from covering its carapace (shell) with a chemically noxious seaweed, which it attaches to hooks projecting upwards. The seaweed provides basic camouflage, like the foliage attached to uniforms of Second World War soldiers, and a chemical defence to predatory fish if all else fails. The crab can be noticed only when it moves – a walking seaweed looks a little odd (again I subconsciously bestow a *set* of characters on an organism so that it may be identified). In the southern regions of its geographic home range, in Alabama and North Carolina, the noxious seaweed is common and is applied by the decorator crab to its shell. The crab snubs the other seaweeds found in this region.

In the northern reaches of its territory, in New England, the toxic seaweed is extremely rare, and as a result is not generally associated with a *typical* environmental scene by the native fish predators. The decorator crab responds by covering itself with a variety of other, more characteristically native seaweeds. It rejects chemical protection to avert an alien appearance, which would otherwise cause alarm bells to ring in the visual systems of predators even when the crab is still. When the northernmost decorator crabs were actually offered the noxious seaweed, they still did not choose it over other seaweeds for camouflage.

Maybe this principle of familiarity could explain why the 'blue frog' is not green. Could a blue frog on a green leaf appear so alien to destroy a predator's means of association with a frog? Well, probably not. The shape of the 'blue frog' suggests it endeavours to be inconspicuous.

Shadows can be the curse of camouflage. Objects of most shapes cast shadows in direct sunlight. A green ball on a green leaf will be easily noticed by its shadow – although only the shadow will be seen, that is enough to suggest that something is unusual and worth investigating. And *predators* tend to be attracted to the unusual.

An average beetle, of spherical or cylindrical form, crossing a leaf would again cast an obvious shadow. Many beetles do live on leaves but have evolved to lose their average-beetle-form for this reason. From above they look like average beetles, but from the side or the front they are clearly different – they appear semicircular rather than circular. Their entire body shape is of a hemisphere. This shape, with flat side against the leaf surface, casts no shadow. Structural modifications are required to adjust to such a body shape, such as legs that can tuck neatly underneath, but the commonness of this design is proof of its success.

Frogs are also somewhat spherical. But many, including the 'blue frog', are able to adjust their shape, thanks to their pliant bodies. Resting on a leaf they flatten down their under-surface, tuck their front feet under their chins, pull their front and back legs into the sides of their bodies and fill the remaining gaps with their back feet until they have moulded themselves into a near-perfect hemisphere. The resting tree frog is shadowless.

Camouflage behaviour

To heighten the enigma of the blue frog on a green leaf, and to maintain its camouflaged shape, the frog employs inconspicuous behaviour. It keeps still, perfectly still, for hours at a time. Many predators, including birds of prey, rely on movement to pick out prey from its background where there is nothing else to go by. Remember the kestrel searching for a vole in the Ultraviolet chapter – a brown vole against a brown background – where, after finding an ultraviolet-marked trail, *changes* in individual pixel colour were the only indicators of the vole's body? The pixels, or rather cone cells, of the retina were all sending messages of 'brown' to the brain, but when just a single pixel changed its shade of brown, it caused alarm bells to ring. The difference in hue was not to blame for this; it was purely that *something changed*. *What* changed, exactly, was irrelevant.

Anything that moves causes some change on a predator's retina, which in turn alters the signals sent to the brain. So clearly life is easier for prey in windy conditions, with so many distractions in the predator's scene. But even now the prey animal must be adapted. In windy conditions a *still* animal will stand out among moving leaves. Since leaves appear as a wave of green flowing across the predator's retina, then so must insects. As a result stick insects and leaf insects have evolved a swaying behaviour. As they move they rock and sway in the manner of a twig or leaf in the wind – in *exactly* the manner; the accuracy of their mimicry is a matter of life and death. And evolution has produced the same result, independently, in the sea. The leafy sea dragon (a fish) appears like seaweed not only in form, but also in movement. The 'dragon' must drift in the water like weed and resist the

Figure 5.2   A tree frog in hemispherical posture to avoid casting a shadow.

urge to dart away when approached by a predator. But this all considers that the animal camouflaged is the prey.

Praying mantids have evolved the shape and movement of leaves too. Perhaps the best example is the orchid mantis of Malaysia. This insect blends perfectly with its lair – the petals of a pink orchid. Both its pink body and pink legs are petal-shaped and sway like the orchid's petals. The only difference is encroachment. While swaying, the mantis ratchets gradually along a stem towards its prey. The sideways movements maintain its camouflage, adding one of those necessary characters to the prey's combination to identify a flower. A movement too rapid towards the prey would appear obviously unnatural – 'something wrong' – and the prey would disappear. Otherwise the mantis reaches striking distance and within a twentieth of a second impales its victim on grasping, saw-edged forelegs.

Movement is the downfall of many paintings. How to portray movement in a still image is a challenge. Animals can avoid this problem – as leaves sway so do the insects mimicking them. But leaves painted on canvas *can't* swing from side to side. They must be captured mid-sway; the problem being that they don't hold still long enough to be painted. But like all things in art, there has been an improvement through time.

On the walls of the Palazzo Pubblico in Siena, near Florence, there is a famous painting of a horse and rider by the fourteenth-century Italian artist Simone Martini. In its time this fresco, named *Guidoriccio da Fogliano*, was revolutionary. Post ancient-civilisation-art was just developing in skill and appreciation. Simone Martini was one of the earliest heroes of the new revolution that had been triggered by the spice trade centred on Florence. Spices quickly became artists' pigments. Florence became a Mecca for artists.

The horse in *Guidoriccio da Fogliano* may have been revolutionary in its time, but appears quite amateurish today – it represents only a first step towards realism (although it does break down boundaries in colour usage). As do all frescos of the Palazzo Pubblico, it lacks perspective – a later addition to art. But the horse simply appears frozen, rather than in the midst of its stride. One cannot imagine that it was really walking, and instead it emerges as a stuffed museum specimen. Once again alarm bells ring in our brain indicating that something is wrong.

Along with perspective, artists subsequently targeted movement. In George Stubbs's *Whistlejacket* the equine subject does portray movement remarkably well, despite the trick played with its background. It could pass for a still from a film of a living horse, as it rears up on to its hind legs. The joints in its front and rear legs flex naturally, its eyes gaze downwards and its blood vessels dilate and protrude from its neck, just as one would expect. Additionally the horse's eyelids distend to reveal the whites of its eyes and portray a frightened disposition, again synonymous with a rearing-up behaviour. All the ingredients of a rearing horse are there except, purposely, for the background. This painting, nonetheless, must have taken weeks to complete, using only sketches and memory since photographic aids lay in the future.

Moving from the eighteenth to the nineteenth century, the French

(Dutch-born) artist Constantin Guys discovered how to portray moving horses on paper swiftly. In 1848 Guys secured a position as illustrator for the *Illustrated London News*, a reputable newspaper. During the 1840s, the issues of the *Illustrated London News* were full of reports of war; in India, Algeria and Mexico. In 1853 Guys was commissioned to record the events of the Crimean War, as a form of pre-camera photographer. This was the first modern war, in that it involved telegraphic communication, a supply railway, and efficient nursing and field kitchens. Guys would sketch representative scenes, which became converted to wood engravings, ripe for the printing presses. He set sail to the Crimea, in the Black Sea, knowing that he was up against ever-difficult horses, hundreds of them, and that to provide a flavour of a battle he must somehow capture their movements in just a few strokes of his pencil. The beginning of the Impressionist movement in art followed.

The bloodthirsty, swashbuckling battles of the Crimean War were preceded by periods of tactical adjustment, in the eyes of the military commanders, or boredom in those of the soldiers. Before a battle, opposing armies would line up parallel to, but some distance from each other. This gave Guys time to choose his spot, overlooking the anticipated battle scene while avoiding danger. And somehow he succeeded in photographing the action with his pencil – today Guys' drawings inject life into the dim record of a time when adventurous soldiers went to battle in brilliant uniforms, armed with swords and rifles.

As horsemen charged, with blades aloft, hooves carved circles into the dusty air like those of the supply-wagon wheels. From where Guys sat, some fifty metres or so from the clash of armies, the straining of individual muscles and looks in the horses' eyes were not apparent. And Guys quickly realised what characters under the 'horse' entry in his visual dictionary were being used to identify the galloping horses at this distance. He realised it was the patterns of movement – those circles created by hooves as well as the arching of backs and swishing of tails. He also realised that the accuracy of some anatomical features was less important, particularly the breadth of bodies and legs, and that this was even more so at distance (as a body moves further into the distance it becomes first a stick-man and eventually a spot). Simply, to look like something else you don't need *all* its characters. The characters you *do* need, though, vary with circumstance, including distance.

Guys found that the best method for capturing movement was to note the position in mid-air of various points along the horse's legs and body. He marked the hooves as simple triangles and knees as circles, and joined them up using just a single, although thick, line. Then he concentrated on the large, upper leg (extensor carpus and triceps) muscles that bulged from the horse's body, and connected these to the circular knees by an elongated, isosceles triangle. This narrative must render impressions of a child's drawing, but in fact movement had presented the most visually important characters for a horse at distance. By using simple lines in just the right places, Guys could capture the realism of a scene better than the detail of a close-up could afford. I have never encountered a more realistic portrayal of horses than those of Constantin Guys' hand, and what separates these from the horses of other artists is the slenderness of the bodies and legs. The exaggerated fineness of his horses' legs, and consequent swollen kneecaps, appear unexpectedly realistic. Slender bodies, in full stretch, infer athleticism and as a result flowing, hurried, *natural* movements. Certainly, Guys' pictures captivate the eye today, bringing to life horses, men and events obscured long ago in the dry accounts of history.

Constantin Guys passed on his methods for capturing movement to

Figure 5.3   Sketch of a Crimean War battle (possibly the charge of the Light Brigade), 1854, by Constantin Guys.

his pupils, Edouard Manet and Henri de Toulouse-Lautrec, methods evident in a number of their works. He had taught the world that accuracy of depiction and realism could be unrelated. Although he did not spell it out in this way, the first and third element in the process of art – the actual scene and the picture on paper – could be different due to the middle process. The visual system employs a certain, counter-intuitive method of interpretation, in which certain things can appear more realistic to the brain if they don't precisely match their models. Famously, as I will discuss a little later, the Impressionists grasped this opportunity with both hands.

This discussion of Guys' art helps clear our thoughts over visual processing. Although the eye itself may faithfully record the world as illuminated by sunlight, the visual system as a whole is rather traitorous. This is because the visual system is the product of evolution. It has evolved to see *certain* things, not everything equally. Faithful images of a countryside scene may fall on to the retina, only because image formation either happens faithfully or not at all. But a complete comprehension of all that data supplied asks far too much from a brain small enough to fit into a skull, and so corners were cut during brain evolution. The 'target' objects that the visual cortex has specifically evolved to interpret are interpreted well; others may be problematic. Guys placed 'target' objects in bold, leaving other objects blurry. But Guys' task was made easier by his obligation to supply *black and white* images.

Camouflage colour

The problem posed by hue is somewhat different to that of behaviour because the images collected by the eye are in colour, yet there is no colour in a real countryside scene. This may seem a particularly high hurdle, but is lowered by a simple solution. Yes, the painting of an orange must appear orange to our eye, and a healthy leaf should be green. But no problem. This is achieved easily by the artist, and by nature too – simply employ the *same* pigments as found in orange peel and green leaves. Then who cares if colour is something added only in the eye? Well, that's fine for most pigments, although structural colours pose more problems, as Monet discovered as he observed a pheasant among his subjects.

Feather-stars ('crinoids') are relatives of starfish. They have long, thin arms in multiples of five, each frilled by short, comb-like branches, and they are often brightly coloured. During dives in tropical seas I have observed many red, yellow, purple and black-and-white-striped feather-stars. I have learnt that these sessile animals are worth a close inspection, since they often host small squat-lobsters ('galatheids') – crustaceans related to lobsters. Squat-lobsters have short, stocky, shrimp-like bodies with long, thin claws projecting ahead. Those found on feather-stars are perhaps the size of a bumble-bee, but are interesting for their colours. They are always *exactly* the same colour as their feather-star host. A strikingly bright hue. Small squat-lobsters employ camouflage colours to hide from both predators and prey. They escape the attention of predatory fish and surprise any smaller crustacean that stops to try its own luck on the feather-star. Indeed the accuracy of their colour matching is remarkable.

When it comes to colour, squat-lobsters do not cheat. They do not extract the pigments from a feather-star to colour their own shells. The existence of a *striped* squat-lobster stifles that idea – this has a body that is striped in black-and-white with the precision found on its feather-star host. Through a little adaptation of its behaviour it can orientate itself so as to disappear from sight. But a squat-lobster certainly could not extract a *pattern* from its host, just as one cannot buy a tin of stripy paint. I am less surprised that bright yellow squat-lobsters have evolved on bright yellow feather-stars, but black-and-white stripes ask a lot more of evolution. Nevertheless, in all squat-lobster cases evolution has succeeded in producing an exact matching pigment, in fact a very similar chemical pigment exists in both animals. In the case of the striped species, where two pigments and a pattern are involved, evolution just takes longer.

Next to feather-stars, on rocks on the sea floor, one may find abalone shells. Abalone are molluscs related to limpets, although they can grow to the size of a saucer. They are popular for two reasons – their bodies are a gastronomic delicacy and they are protected by scintillating, iridescent shells. Maori and other Polynesian people incorporated abalone and related paua shells into their jewellery and carvings. But abalone and paua do not appear so attractive when alive. In fact they are difficult to see at all.

Behind the abalone shell's beauty is a multilayer reflector. Part of the shell is composed of thin layers, which evolved this way for reasons of strength, in the manner of laminate wood, and to prevent crack propagation (a crack will spread through an entire thick layer but will terminate at the boundary between two thinner layers). As luck would have it, the thin layers in the abalone shell are around a ten-thousandth of a millimetre in thickness, although only because this is perhaps the most convenient size for the abalone to construct, given the limited processes of manufacture at their disposal as they evolved. But this is the magic number for multilayer reflection. Since the layers are also transparent they do indeed cause an incidental structural colour – that seen in Maori jewellery. So, then, why are living abalone difficult to find?

I specified that abalone iridescence is incidental. True. Abalone possess neither eyes nor toxic defences, making bright warning colours as futile for them as iridescent mating displays (in any case, abalone do not mate but instead spawn their reproductive products into the ocean, where fertilisation takes place). Simply, bright colours could serve no purpose for the abalone. They could, and would, on the other hand, be disadvantageous to the abalone – they would alert predators to their whereabouts. For these reasons their behavioural balance has tipped in favour of inconspicuousness and the abalone is camouflaged. So during abalone evolution, the iridescent part of their shells must have been either altered, so the layers became too thick or thin to comprise a multilayer reflector, or concealed. In their case, a pigment evolved to mask the rebellious iridescence. This represents the simplest solution but also one with a dual purpose. The pigment that has evolved as a barrier to iridescence is pale brown, the colour of its rocky background.

The pigmented hues of squat-lobsters and abalone are no-nonsense cases of camouflage coloration in its simplest form, of which there are countless others. But camouflage coloration can be more elaborate, as in the case of the magpie lark on lily pads.

The magpie lark is an ancient, Australian bird, about the size of a miner bird and pied like a panda. It can be found in many situations, such as among the lily pads of ponds where it hunts for dragonflies and damselflies. As everyone knows, lily pads are green, but unlike oak or beech leaves, for instance, they have shiny surfaces. The waxy coating of lily pads serves to repel water – raindrops will roll off just as pond water will fail to permeate. The surface of the wax layer is by its nature extremely smooth at a microscopic scale, and smooth surfaces also reflect light, particularly when the light approaches at a glancing angle.

Where lily pads are packed into a pond, their competition for space results in leaves pushed partially into the air. Lily pads are often found half on the water and half folded upwards. An onlooker will see sunlight reflected strongly from parts of each leaf, while other parts are orientated to reflect light into other directions, like a series of mirrors set at varying inclines. Where a reflection is not seen, the green of the leaf is evident, although appearing rather dark at a distance. So the lily pads emerge to the onlooker as patches of 'black' and white (shaded regions add further black patches). Regardless that we associate lily pads with green, for animals that reside on their surface black and white patches offer the most suitable camouflage coloration. And that is the response of evolution in the case of the magpie lark. The selection pressures acting on this bird are the colours as they appear on retinas in the *natural* habitat only. Evolution does not work in the round-about, thought-oriented way that *we* have approached this problem, beginning with green leaves. But it does work successfully – the magpie lark is near impossible to spot on Australian lily pads, as it is in other environments where the bright, Australian sunlight becomes reflected.

Although the pied coloration of the magpie lark offered direct camouflage coloration against a natural background of lily pads, in most other environments its black and white patches provide disruptive coloration.

Disruptive coloration is where patches or patterns of different colours serve to break up the outline of an object. The stripes of a tiger or clown fish and brown blocks on a giraffe's hide achieve this. Modern army clothing or camouflage-painted planes are other good examples.

Figure 5.4   The black and white magpie lark (centre), camouflaged against lily pads.

Military planes may be painted in blocks of greens, yellows and pale browns – the colours as farmers' fields. Viewed from above, this plane will merge into those fields very effectively, but not in the same way as simple camouflage coloration. Disruptive coloration works on the principle that some part of an object will be visible, but others will not. The shape seen becomes unrepresentative of the object itself, and the object is no longer identified.

The reason why the shape of an object is destroyed by disruptive coloration is that the eye does not see in outlines, only in pixels. This characteristic emerged in the Ultraviolet chapter, where the kestrel did not detect the shape of a brown vole against the various shades of brown in its background, but rather reacted to changes in individual pixels within the whole scene. The vole employed a uniform camouflage colour, so only when its entire, immediate background lacked that hue was the shape of a vole formed on the retina.

Disruptive coloration involves risk-taking. Its methods accept that part of an animal will be visible, but the *range* of colours offered for camouflage avoids resting all hopes on one. As the military plane passes over fields, at one instant the brown patches will overlie a green field and will be obvious, but a yellow patch will overlie a yellow field and become camouflaged. The result is a messy image on the retina, where

Figure 5.5   Military plane in camouflage (reproduced with permission from the Imperial War Museum)

patterns are random and without the shape of a plane. A plane is not identified in the brain's dictionary.

Many moths employ disruptive coloration as a means of appearing inconspicuous against patchy, inconsistent backgrounds. Usually this works – most birds capture moths only on the wing, since they simply can't find them against mottled bark. The exception is the blue jay of North America. Researchers from the University of Nebraska at Lincoln, USA, noted that this bird actively seeks out moths while they rest against bark. Through recent experiments they deduced that the blue jays' visual dictionary contains more than one entry for moth. It also contains certain patterns found on the moths' wings, such as a pair of white dots or a black 'M' shape. Although their colours are seen commonly on bark, the patterns themselves are uncommon, and the moths become visible to the blue jay (although not to humans and other birds). Of course evolution marches on and the moths have evolved to become more variable within a population, presenting a variety of patterns on their wings rather than always white dots or black 'M's.

The brain struggles less with black and white disruptive images. The famous illustration to back up this statement is the Dalmatian dog pictured against a background of random black and white patches. The outline of the dog's body is absent, and its spots merge with the black spots in the background. Still, most people can pick out the image of the dog. That's because the information received by the brain is in binary form – groups of neighbouring red, green and blue cone cells in the retina are either all 'on' or all 'off'. In white patches they all fire signals, in black patches none fire. But add colour to the pattern and the whole image becomes more complex. And the more elaborate the data, the more problems created in the brain. This idea was exploited by Cubist artists, who played on the edge of recognition and unawareness of objects constructed from shapes, as the boundaries of those shapes disappear at crucial places.

A central problem is that although the retina collects visual information from a natural scene as a series of pixels, in the manner of a digital camera, we do not *see* in pixels. We do not detect our visual world as an array of coloured dots. We see birds, trees and faces as continuous, unpixelated surfaces. So somewhere in the brain the pixelated information supplied by the retina becomes re-encoded. The objects we finish with in our mind's eye possess smooth, continuous

Figure 5.6   Can you find the Dalmatian? (From Thurston, J. and Carraher, R.G., *Optical Illusions and the Visual Arts*, Litton Educational Publishing Inc., 1966)

edges and surfaces. So the yellow patch on the military plane merges with the yellow field it overlaps below to become one large, smooth patch of the same, continuous hue.

This is where the Impressionist artists, with their style of 'pointillism', further manipulated our visual system. We will always end with an image of smoothness, and always pass through an intermediate stage of pixelation. So it matters not whether the scene we view is in smooth, continuous form or in pixelated form. But a pixel on a Pissarro painting can fall on the border of two cone cells, and so resonate between them. In this way a pixel can in one instant belong to one object, and in the next instant belong to another, juxtaposed object. The result is a sense of movement or realism – the static of a Constable landscape is broken. My lengthy description of the seeing process, as followed by NanoCam in the Introduction chapter, suddenly seems worthwhile, as now we are equipped to make sense of all manner of artistic traits.

Once more we find ourselves in accord with the concept of 'seeing is seeing colours', first introduced in the Ultraviolet chapter. Here, the discrimination of different wavelengths of light is considered at the core of vision. In which case, 'seeing colours' would appear the general selection pressure for the evolution of animal colours, not to mention art.

Evolution of camouflage colours

Nineteenth-century biologists toyed with an early theory for the evolution of colour pigments. Photography was a new trend and held the fascination of all scientists. Accordingly the principles of colour plates or film were shoehorned into the subject of animal colours. 'If it was possible to obtain *inorganic* chemicals which are so extraordinarily sensitive to light,' thought the Victorian biologists, referring to photographic compounds, 'it is surely not impossible that *organic* substances, in their ordinary position within the organism, may display a similar sensitiveness, and therefore that pigment production may be the result of exposure to light.'

The Victorian idea of camouflage accomplishment – that animal

skins contained chemicals for photographing their environments –
attracted further interest when a method for *colour* photography was
suggested. Now animals were thought to possess 'substances which
react in such a manner to different rays of light as to themselves build
up compounds having the same colour as the incident light'. Certain
compounds of silver chloride were known to achieve this, and the biologists extrapolated that organic substances held the same property,
accounting for camouflage colours. 'A caterpillar may be like its environment, because its skin photographs that environment by means of
the sensitive compounds of its own tissues,' according to Herr Otto
Wiener. Of course Wiener was wrong.

By the end of the nineteenth century many studies had confirmed the
relationship between the colour of a pigment and its chemical composition. This led to new ideas on the evolution of animal colours.

Another Victorian biologist, Dr Urech, had studied the development
of some butterflies (of the genus *Vanessa*). He noticed that the wings
began as white, but then 'develop in the order of the spectrum (yellow,
orange, red, brown, black)'. This, thought Urech, 'suggests that there is
a relation between the molecular weight of these pigments and their
respective colours, and that this gradual development of colour in the
history of the individual corresponds to the evolution of colour in the
history of the race'. Fellow biologists had found supporting evidence of
Urech's theory in the purple pigment 'rhodopsin', which was observed
to change through red, orange, yellow until finally it became colourless.
The revised hypothesis for colour pigment evolution was that all animals contained 'simple' chemicals in the cytoplasm of their cells, of
ancient origin. These chemicals could first become black and then, as
the molecules became increasingly complex, brown, red, orange,
yellow, green, blue and violet, in that order (presumably followed by
ultraviolet, had it been known in animals at that time). So achievement
of a violet pigment would involve the most activity, having passed
through a complete spectrum and several intermediate compounds *en
route*. Accordingly, also because it was predicted as the largest pigment,
it came at the highest energy cost to an animal. This idea would explain
the greater prevalence of reds, oranges and yellows in nature, but twentieth-century chemistry would topple this house of cards.

There is no pecking order in animal pigments, and no general reason why the pigment for one colour should be more difficult to evolve than that for another colour. That's due to the real mechanism for pigment evolution – all pigments have their evolutionary origins in a chemical reaction not related to visual adaptation.

Colour pigments come in a variety of chemical compounds. Chromium compounds produce yellow effects, potassium compounds blues and copper compounds greens, to name a few. They arise originally from chemical reactions that take place in different parts of an animal, as part of the general running and maintenance of the body. Where these reactions take place internally, the colour of the chemical products is unimportant. The iron in blood imparts a red colour, but its purpose is *only* to collect oxygen molecules. The red colour is immaterial, since it is not observed from outside the body (nor is the blood naturally reached by sunlight to fuel the red colour). For animals that live underground, no light ever reaches any part of their bodies so their colours are all incidental (I avoid using the word 'accidental' since everything in nature is accidental, that's the way evolution works) – they are observed only in unnatural situations. They do, nonetheless, often possess colour, and that is important to colour adaptation.

Some cave or soil-dwelling animals are white, in that they possess no colour pigments. But most do appear coloured when they are unearthed. Tube-dwelling marine worms often appear orange; some earthworms appear pink. In these cases the colours are never fuelled in their natural home, but the fact that the colours exist indicates that they do arise incidentally. Now if that tube-dwelling worm, throughout geological time, inches out of its tube, it may evolve protective spines or toxic chemicals in response to the new selection pressures posed by the new threat of predation. Its new predators will include animals with eyes, and so selection pressures will act also to adapt the worm's visual appearance. The orange pigments would become the starting point for the evolution of adaptive colour. They may become brighter, to warn of the spines and toxins, or they may become duller to provide camouflage against a sandy sea floor. In fact both scenarios will take place within a population of the worms due to genetic mutations; the scenario within the worms that survive best will be that inherited (since those worms will, through avoiding predation, exist in greater numbers to pass on their genes for *their* colour).

So a pigment that begins its evolutionary life as incidental can become subjected to natural and sexual selection, to increase its host's life expectancy and mating potential. And that's how all pigmented colours we see in animals today came to be. Where the coloured product of a chemical reaction exists in an animal's *skin* – the part of the animal exposed to the environment – it must become adapted to sight through evolution. As demonstrated by the abalone, a colour that is incidental and maladaptive can be dangerous. But sometimes a chemical starting point for the evolution of a camouflaging pigment is never reached. This is a common problem for animals living on leaves – the raw materials or foundations for green-pigment-evolution must be rare, since there are surprisingly few cases of green pigments in the animal world.

Camouflage brightness

There is a little more to the idea that 'seeing is seeing colours'. A coloured appearance not only boasts a precise hue but also a certain *brightness*. Leaves are generally green and distinctly dull – the brightness associated with a pigment (owing to the dilution of the sun's rays through thin-spreading into all directions). In contrast multilayer reflectors, often employed by beetles as a substitute for green pigments, appear characteristically bright (since the sun's rays are concentrated into one direction). In other words, a beetle with a multilayer reflector will not match the brightness of a pigmented leaf. Not only will it appear green from just one direction, but even if viewed from this direction the green colour will appear too bright. The green cone cells in a predator's retina will fire slowly where they view the leaf, and rapidly where they view the beetle. Fortunately evolution has found a solution to the otherwise dazzling beetle – on top of a multilayer reflector the beetle has evolved a scattering structure.

The very outer layer of the green beetle's armour has become crumpled. It causes sunlight that strikes the beetle to bend so that only sunlight approaching from one direction – precisely that for green – makes it through to the multilayer reflector. That disposes of the blue and yellow reflections. Then the crumpled surface acts on the reflected green rays on their way out of the beetle in the manner that a pearly, 'diffuser' coating affects a light bulb. The crumpled surface bends the green rays randomly, diverting them into all directions so that collectively they cover the entire hemisphere. So the otherwise bright, green reflection is spread evenly over all angles, appearing dull when viewed from any position, just like the appearance of a leaf. Now the beetle has achieved camouflage success – a coloured appearance with the hue *and* brightness of a leaf. And brightness becomes even more critical to camouflage when the observer sees only in black and white.

Camouflage in black and white

Humans are atypical of mammals in that they possess three cone cells – red, green and blue. Most mammals, including nearly all monkeys of South America, have just two cone cells in their retinas, those for green and blue. They cannot distinguish between the colour of red fruit and green leaves. More importantly, they cannot distinguish between a green, unripe fruit, and a red, ripe fruit (and also old, green leaves from young, more nutritional, red-tinged leaves). A fact that, on its own, offers an explanation for the evolution of full ('tri-chromatic') colour vision in mammals.

The evolution of the red cone took place in monkeys. Monkeys of Africa and Asia, as well as apes, all possess the red cone, suggesting that the red cone evolved before the apes. Many marine mammals such as dolphins and seals, on the other hand, moved in the other evolutionary direction. They lost their blue cones, leaving only green and, as a result, 'black and white' vision. The most recent theory holds that when mammals re-entered the water they occupied only shallow regions. Here, blue rays were scattered and removed from the usable sunlight (this part of the theory should be improved), and blue cones became surplus. The blue cones were lost but when the mammals later moved to deeper waters, where blue rays actually dominate the sunlight, the evolutionary process could not be reversed. The marine mammals could not evolve blue cone cells. That a character lost in the evolution of an animal group cannot re-evolve further down the evolutionary line is common. For it to re-evolve, evolution must be directly reversed not only to the point where the character was lost, but to the point where it first evolved and the raw materials existed on which selection pressures may act. Those raw materials do not exist in the living representatives of an animal group for ever as the group continues to evolve. In the case of marine mammals and cone cells, evolution must be reversed an unfeasibly *long* way.

Such 'monochrome' vision is employed also in some other animal groups including, surprisingly, the cephalopod molluscs – octopuses, cuttlefishes and squids. Cephalopods, as I will discuss in the Orange chapter, have evolved an array of colour-producing mechanisms, and so appear highly coloured. They just can't see their colours themselves, although obviously other animals can.

Marine mammals and cephalopods themselves are the main predators of, in particular, cuttlefishes and squids. That would explain why some cuttlefishes, while resting, motionless on the sea floor, appear quite obvious to humans. They are camouflaged only for the benefit of their predators with black and white vision. Remember, red and green appear similar to an animal without red cones, so camouflage against a green background can be achieved using red pigments. That's very convenient, since cephalopods have only brown, red, orange and yellow pigments at their disposal – only the starting points for the evolution of those pigments existed in the body chemistry of cephalopods.

So could *this* explain why a blue frog should choose to rest on a green leaf? Could blue and green appear equal to an animal with black and white vision? Well, no. Not only do blue and green appear different in black and white, but the main predators of the 'blue frog' would include birds and reptiles, both of which have red, green and blue cone cells (and in the case of birds an extra cone too). They see in full colour.

The next wave of zoological study in Australia

Winding back the clock 150 years from today, zoological research had just received quite a boost. Darwin and Wallace had just published their theories of evolution, Henry Bates had exposed mimicry, and natural history in general had become trendy with the general public. But the more casually inclined 'naturalists' had become strictly 'scientists', who retraced earlier footsteps into the forests of New South Wales and the territory of the 'blue frog'.

Before long another 'blue frog' was encountered in the wild, this time by a new-model scientist. At first it was unclear what had been found – a new species must have been assumed. The frog peeled from a leaf was not found easily. It was not blue but green. But as descriptions of the discovery commenced back in the office, problems arose. The features listed – the shape of the sucker-like feet, the size of the body, the webbing of the hind feet only – had all been expressed before. All except one, that is – the colour. The description matched that of *Litoria caerulea* – the '*blue* frog'.

It was clear that the new tree frog had to be officially designated *Litoria caerulea*, the name to be added to its paper label in its alcohol-filled storage jar. White's 'blue frog' had surfaced as the green tree frog!

We seem to be getting somewhere. The problem at the beginning of this chapter is now looking suspicious. It would appear that the *living* tree frog seeking cover on an Australian green leaf was itself green. So why did White's tree frog appear blue when it was lifted from its rum bottle? And, for that matter, why do specimens of *Litoria caerulea*, packed into glass jars gathering dust in museum annexes, appear blue today in their alcohol preservative? Time, eventually, for NanoCam.

Getting under the tree frog's skin – detail of the colour factory

Now it all makes sense. Out in the forests of New South Wales, Australia, a living specimen of *Litoria caerulea* is observed with difficulty.

It is green, curled up into a hemisphere, motionless and casts no shadow on to the green leaf it has chosen as its background. It could easily pass for a rubber toy frog. Indeed, I was fooled on my original encounter with this species. While travelling with friends along the east Australian coast, touring the beach hostels joined up by Highway 1 (the single road that circumnavigates Australia), I once opened the cutlery drawer in a rather humid and accordingly relaxed kitchen one afternoon. Knives were visible in one compartment, forks in another, but the spoons were obscured by a 'latex' frog. I laughed as I wondered who was responsible for this practical joke, right up to the point where I picked up the frog to access a spoon and it jumped out of my hand. Certainly, the green tree frog is good at keeping still, and is difficult to find among a forest canopy.

Returning to the newly found frog on its leaf, NanoCam's probe is edged towards it, approaching cautiously from behind. On making contact with the green skin, the camera optic manoeuvres between spaces within a layer of dead but transparent surface cells and is injected into the cells beneath. Two types of cells appear common, arranged in near-distinct layers.

NanoCam punctures a cell membrane of the first cell type to be encountered. Immediately evident is the lack of electrons jumping between orbitals within the cell's molecules as they meet light rays – pigments must be absent. There are, nonetheless, light rays bouncing around within the cell, heading in all directions, which is unusual since light enters from only one direction. But not only are pigments absent, there are also no signs of multilayer reflectors or diffraction gratings, those optical devices that cause iridescence, nor bioluminescent reactions. Yet something optical is happening. NanoCam scans the cell to find out what.

The watery fluid 'cytoplasm' inflates the cell and holds the cell membrane rigid. Within this watery matrix lie the cell's nucleus and organelles, where the activity of the cell takes place. These are imaged clearly by NanoCam due to their sheer size in the context of the cell. But after some time swimming through the cytoplasm NanoCam begins to take seriously the 'debris' that keeps lying in its path.

The cytoplasm is littered with tiny particles, arranged completely at random, that perform no apparent function and seemingly contribute

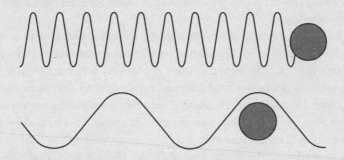

Figure 5.7 A short wavelength is stopped or reflected while a long wavelength is not affected by the same particle.

nothing to the cell. Yet they transform the otherwise clear cytoplasm into what looks like a dust cloud on the NanoCam monitor. Individual particles, or 'dust', vary in size, but all are smaller than 575 nanometres – the wavelength for yellow light. And now that NanoCam has focused upon a group of them, they *do* appear to affect light.

Rays of sunlight head into the dust cloud. The red rays pass almost straight through the cloud and continue along their path towards the body of the frog, although a tiny proportion strike and reflect from dust particles. The blue rays are reflected rather more. The orange, yellow and green rays lie somewhere in between. But we see only the extremes – the reflected light appears blue, the transmitted light red – since rays of these colours dominate the two light paths.

Generally, the longer wavelength red rays straddled the dust particles and continued along their path, while blue rays bumped into the particles and were reflected. NanoCam focuses on those reflections.

One blue ray hits the edge of a dust particle and is reflected sideways. The next blue ray hits a particle head on and is reflected back on itself. In fact blue rays are observed bouncing around inside the cell in all directions, like a pin-ball machine. A similar effect was achieved by the scattering particles in the fire-fly's photophore, as described in the Blue chapter, except there the individual dust particles were larger. The particles were so large as to reflect all wavelengths equally – not even the red rays could dodge the dust 'rocks'. That reflector would appear white in sunlight.

The overall effect of the cytoplasm dust cloud is that blue rays leave the cell equally in all directions, while other colours mainly pass through the cell. The blue coloured appearance is similar to that of a blue pigment in that sunlight is spread evenly into all directions. Unlike a pigment, though, the transmitted rays pass straight through the cell rather than being absorbed and converted to heat. This optical process is known as 'Tyndall scattering'.

In the fifteenth century Leonardo da Vinci hinted that the sky appeared blue due to an optical effect. But confusion over the precise mechanism followed until the Victorian experimentalist John Tyndall revealed the answer in the nineteenth century (an explanation that was further expounded by Lord Rayleigh soon after). Armed with knowledge of the wave nature of light, Tyndall explained that fine particles and water molecules in the Earth's atmosphere reflect the blue rays in sunlight down to the Earth's surface, while the red rays mainly pass through the atmosphere and into space. We view these red rays that pass directly through the atmosphere only when the sun is low on the

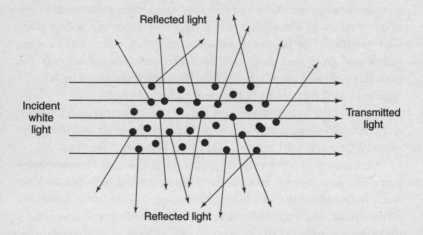

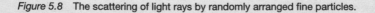

Figure 5.8   The scattering of light rays by randomly arranged fine particles.

horizon, during sunset. That's why objects appear redder at sunset. Uluru (Ayers Rock) in the centre of Australia is the perfect example, changing in colour from yellow at midday to red at sunset. But in the absence of the atmosphere the sky would appear dark during the day, apart from the dazzling sun.

I was fortunate enough to play with John Tyndall's experimental demonstration of blue scattering at the Royal Institution in London. The public lectures here, often with a 500-strong audience, are famous for their demonstrations. Lecturers are encouraged, if not obliged, to flesh out their talks with moving models, flashing lights and the occasional explosion. This has been the trend since Michael Faraday opened the long-running lecture series with demonstrations of *his* discovery – the production of electricity. Some subjects lend themselves better to demonstrations than others.

Also near the origins of the Royal Institution Lectures, Tyndall aimed to show and so convince his audience how fine particles could result in a blue colour. He took a glass box, the size of a small home-aquarium, a lamp with a narrow beam of white light focused on to a white screen, and a long paper 'straw'. He set light to the straw, blew out the flames, opened slightly the lid of the glass box, and allowed smoke from the smouldering paper to fill it. The glass box was positioned so that it intersected the beam of the lamp.

The glass box radiated a blue colour and the white spot on the screen turned to red. John Tyndall had demonstrated that fine particles in a transparent matrix, in this case smoke particles in air, could scatter the blue rays in white light more than the red. In 2000 I used exactly the same equipment and to my astonishment it worked! A beam of white light entered the glass box, the box glowed blue and the spot formed on the white screen by the remaining beam almost condescendingly transformed to red. I had my excuses to the audience well prepared, but the success of the experiment caught me unawares and an unscripted moment of silence followed. At least my 'fluorescent parrot' exhibit failed, but that's a subject for the following chapter.

So, we have a layer of cells in the green tree frog's skin that reflect blue rays via Tyndall scattering. Now NanoCam burrows further into the frog's skin and encounters a new layer, filled with the second type of cell. This cell is obviously different.

The second type of cell contains a molecule that is affected by light. The NanoCam monitor images a single molecule as a strong beam of sunlight strikes the frog. The molecule is instantly identified as a pigment. The crash observed, as light rays collide with the molecule, is followed by unnatural electron movement. Electrons begin to jump between orbitals within the molecule. For this to happen, energy is required – it is extracted as the molecules eat up most of the sun's rays; that is those rays left after Tyndall scattering has taken its toll in the layer of cells above. Eventually that energy is converted to heat – NanoCam warms up. Paying a little more attention, some rays are obviously not consumed by the molecule – they are instead reflected. And the reflected rays all have one thing in common – their wavelength. They all have a wavelength of around 580 nanometres. That's yellow to us. This molecule is a yellow pigment. The cells of this second type are bursting with yellow pigments.

Taking a step back to re-assess the case of the green tree frog as a whole, we should identify the three visual elements in the colour system. The first visual element involves a *green* reflecting leaf. The third element, the animal that has evolved camouflage coloration to disguise itself against the leaf, involves a *blue* and *yellow* reflection. So what does this tell us about the second element, the visual system (eyes and visual cortex) of the target predators?

Colour mixing

The cone cells in a retina each have a response curve. That is, they are better at detecting some wavelengths of light than others, but will detect a range of wavelengths all the same. The wavelength best detected will be at the centre of a response curve that drops off into wavelengths above and below.

Humans, as we have learnt, have three types of cone cells in their retinas – blue, green and red. The blue cone has a peak response at 420 nanometres ('violet'), but tails off completely at around 540 nanometres ('green/yellow') and into the ultraviolet in the shorter wavelengths (although the cornea of our eye absorbs light below 400-nanometre wavelengths). The green cone has a peak response at 530 nanometres ('green'), the red cone at around 565 nanometres ('yellow'), both again tailing off in their response at wavelengths above and below. The combination of all three cones means that we see *all* rays from 400 to 700 nanometres – violet to red. But due to the overlapping ranges of each cone (see Figure 1.2), most wavelengths will be detected by more than one cone.

When describing a colour we do not say 'a large red and green signal and a little blue'. We say 'bright yellow'. And that is of the essence in the case of the green tree frog. Crucially, there is no single cone cell just for 580-nanometre (yellow) light.

An interesting experiment exists to prove this point. Three transparent but coloured squares are positioned on a clear, transparent slide so that two of them partially overlap. One square is coloured with pigments reflecting 580-nanometre wavelengths. This square stands alone and appears yellow. The remaining two squares are coloured, one with a 540-nanometre pigment, appearing green, and the other with a 620-nanometre pigment, appearing red. When the slide is projected on to a white screen using a white light, the area where the green and red squares overlap appears yellow – exactly the same shade of yellow as the isolated square. Where the green and red pigments superimpose, our green and red cone cells are stimulated in just the same way as by a yellow pigment. So 'yellow' can derive from a single wavelength of 580 nanometres, or the combination of 540- and 620-nanometre wavelengths. There are literally both options for yellow.

With this in mind, it may be worthwhile considering the actual wavelengths reflected from *Litoria caerulea*, which appear green to us. Could this frog's perceived colour differ from the direct translation of the rays reflected?

The living green tree frog is now assessed with a spectrometer. Unlike the eye, spectrometers measure *absolute* colour – the real reflectance for each wavelength from 400 to 700 nanometres, for instance. The spectrometer probe collects all the rays reflected from the frog in sunlight, and produces a real spectral curve. The curve is interesting, and maybe not what we would have expected. *Our eyes say green frog; the spectrometer says blue (cyan) and yellow frog*. The spectrometer's spectral curve, the only honest record of the wavelengths leaving the frog's skin, has two peaks – one in the blue region of our spectrum, and one in the yellow. There is no single peak. There is no reflection at all in the green region of our spectrum. As I said, this is not what we would have expected, at least before I described the coloured square experiment and investigated with NanoCam. But in the wake of this experiment and investigation we have our two and two to put together, beginning with how *Litoria caerulea* appears green.

The solution

So the green tree frog is another victim of the illusive evolutionary pathway to green pigments. The chemical reactions in the body of *Litoria caerulea* never offered a surplus product that would become a fully-fledged green pigment via genetic mutations. The frog did, on the other hand, evolve a yellow pigment – that packed into the second cell types in its skin. Those pigments cause the yellow peak in the spectrometer reading.

The blue peak resulted from Tyndall scattering – the pin-ball reflections of the blue rays in the first cell types. Of the green, yellow, orange and red rays that continued their path through these cells, the yellow was reflected and other colours absorbed via conversion to heat by the yellow pigments beneath. Now the fate of all colours in the incoming sunlight has been accounted for.

The yellow reflections are detected by the red and green cones in our retinas; the blue reflections by the blue and green cones. A green reflection would also call all three cones into action, and the result is that a green reflection is interpreted in exactly the same way as a yellow-plus-blue reflection. This offers two options for camouflage against a green leaf – the green tree frog has evolved the yellow-plus-blue option.

We can assume that the predators of the green tree frog – the lizards, snakes and birds – see the frog in the same way as ourselves. They have a similar visual pathway. Most importantly, they have three or four classes of cone cells with which to see colour, each with its own spectral response curve. In other words, the colour of the green leaf and the blue and yellow frog would appear inseparable to the frog's predators – both objects register as green. The frog is successfully camouflaged in colour as well as in its shadowless shape and motionless behaviour. Now all that is left to explain is why John White used the Latin word for 'blue' when he named the green tree frog of New South Wales. A clue towards solving this problem can be found in Egypt.

During an excursion along the River Nile, one encounters many temples showing remarkable preservation. I remember how each temple was so different to the next, but also noticed the colours that had been preserved. The occasional patch of bright pigments amid the sandy-coloured pillars and walls can be as appreciable as the engineering feats themselves. That 3–4,000-year-old paints continue to shine as brightly today as when they were first applied is extraordinary, but it only takes the lithographs of David Roberts to quickly explain how this is possible.

The British artist David Roberts was unfairly forgotten soon after his death in 1864, but was rediscovered recently by art critics. One of his great achievements was his series of drawings of Egyptian monuments, from 1938 to 1939. Roberts is perhaps not so revealing of the nature of vision since he drew with great precision and realism. No optical illusions, technical short cuts or visual trickery of any kind. Nonetheless, although he sketched in grey pencil, he did supply copious notes on the colours of his subjects. These came in useful, as the lithographic prints of his artwork were compiled into 'subscriber editions' of books, where Roberts oversaw the addition of watercolours – *accurate* colours.

David Roberts offers an opportunity to observe a complete temple in one view; during a visit to a temple we spend most of our time staring at ceilings or with our face pressed against a wall of hieroglyphics, or otherwise semi-blinded by the sun's intense glare. Roberts's overviews provide an idea of both colour and scale combined, and so the areas of colour preservation can be placed in context.

Roberts's depictions of the Temple of Isis on Philae reveal columns and walls supporting a carved roof 'all in excellent condition, with brilliant colours', as he records in his journal. I remember approaching the tiny island of Philae in the Nile, today near to the Aswan Dam, by boat in the early (pre-tourist) morning. I recall the vastness of the temple and its elements-defying condition, apparent even from the water. Next I remember gazing up at the colours on the ceilings – David Roberts was right. But looking at Roberts's coloured lithographs of this temple it is clear that the colours exist only on the ceilings and the tops of the columns. Colour is not apparent where the sun burns strongly into the sandstone, but only where relief is sought in the shade. And, as one would expect but possibly neglect, the brightest colour lies in permanent shade, while some colour exists where lit directly by the sun for part of the day, and those parts of the columns heavily flayed by the sun are down to their bare sandstone. In other words, sunlight (particularly the high-energy ultraviolet component) fades pigments – the chemicals are broken down. But the structures – the columns and even the tiniest hieroglyphics carved into them – have been preserved.

Now take the green tree frog, *Litoria caerulea*. At this point we know that its colour results from the combination of a yellow pigment and blue Tyndall scattering. These colour mechanisms can be compared to the paints on, and carvings in, the Temple of Isis at Philae. The frog's yellow pigments and the paints are similar. But so are the carvings and the frog's Tyndall scattering system – the scattering system is a *structure*; the frog's blue is a structural colour. If a dead frog were left to dry out in the sun, we would expect its yellow pigments to fade like those of the temple's paints. And we would not expect the Tyndall scattering system to last very long either, since the fine particles in the cells are held apart by the watery cytoplasm. The cytoplasm would soon dry out, leaving the particles to collapse and scattering to terminate – the structure would fall. But consider the preservation of John White's frog, not in the sun but in alcohol.

Pigments fall from a paintbrush as it is dipped in white spirit – alcohol, like the sun, destroys pigments. And so were the frog's yellow pigments destroyed by the rum, even causing a discoloration. The skin cells, nonetheless, did not dry out. They lost their cytoplasm, but gained a new watery matrix – alcohol. And alcohol preserves the cells. It fixes the fine particles in the first skin cell types in their natural positions.

The preserved frog lost its yellow pigments but held its Tyndall scattering structures. The preserved frog became blue. Today, specimens of *Litoria caerulea* preserved in alcohol (ethanol) in museum collections are also blue. We have our answer.

A tonic for Darwin

This chapter carries the fourth lesson that the eye is not perfect. A predator that strives to catch a green tree frog has great difficulty in finding one. The frog is actually incredibly obvious, shining out in blue and yellow amid a canopy of green. But the retina cannot distinguish between a blue-plus-yellow reflection and a green reflection. Due to the design of the retina, a bird or reptile predator cannot see its obvious food. The eye is not perfect when it comes to colour vision. Again, Darwin did not need to worry about the eye being *too perfect* to result from evolution – it is not. The optical trickery covered in this chapter highlights another fault in the visual system, although this will be the theme of, and expanded in, the Orange chapter.

We will remain in the canopies of forests to consider parrots of Australia, Africa and South America in the next chapter. Daylight and moonlight will provide interesting comparisons, and birds will be exploited to divulge the extremes of behavioural strategies – camouflage and conspicuousness. Flight has handed carte blanche to birds when it comes to colour, and some parrots have responded in a way that at first seems beyond possibilities. A new mechanism for colour production will emerge.

COLOUR 5

yellow

The problem:
Why do some parrot feathers appear bright yellow to us when they
are lit by ultraviolet light only?

Otto Völker was a German pathologist at the Kaiser Wilhelm Institute
for Medical Research in Heidelberg. In 1937 he published an eleven-
page paper in the *Journal of Ornithology* in which an unusual colour
effect was reported in birds. It should have created a stir in scientific
circles, but didn't. This was the era of physiology, with shiny new
machines providing novel views of cells and breakthroughs in molecular biology to go with them. Advances in *colour* studies belonged to the
previous scientific age.

Völker's paper was written in German and was probably read little
if at all outside Germany. It doubtless suffered the fate of many non-
physiological papers of that time, being stored in the 'curiosity' file.
Even in its time it failed to make the reference lists of other ornithologists, and when that happens, you're really doomed. Völker's work
became, simply, lost. Lost in the archives, that is.

On the other side of the Earth, André Nemésio was just fourteen
when he raised the 'Völker 1937' paper from the dead. That it should
rise again in Brazil would seem most unlikely, were it not for one link –
parrots.

Falling into safe hands

Belo Horizonte is the third largest city in Brazil, the capital of the state of Minas Gerais, lying just north of São Paulo. It is modern, springing from an architect's drawing board, and has three universities and breathtaking National Parks on its doorstep. Belo Horizonte lies on the edge of Brazil's tropical rain forest, and it was here that André Nemésio developed his passion for parrots.

'You can easily hear them,' was André's first comment when asked whether parrots are easy to find. Although a couple of species are common flying in between the city buildings, amid the dense forest canopies 'several species are easily seen'. 'If you go to the correct places, you can see large groups of macaws nesting,' explained André on a subject of classic, natural history film footage. Although he also introduced a note of caution for bird-watchers, with his comment 'some other species are hard to find' – shy parrots also exist in Brazil.

Birds have an obvious edge in nature's arms race – when danger approaches, they can take to the air. With sharp eyesight for sensing a predator's approach (thanks to eyes that are larger than their brains), they have often evolved along the road to conspicuousness – bright colours are a considerable plus when it comes to mating. And a successful mating strategy can ensure the survival of one's species – evolutionary safety in numbers. In many cases, parrots are no exceptions.

Parrot diversity may be a consequence of fruit and nut evolution. In the previous chapter I suggested that primates evolved colour vision to distinguish between ripe (red) and unripe (green) fruits. Well, fruits themselves then evolved to become larger and more colourful in response to the new selection pressures of seed-dispersers with full colour vision to complete the cycle. And parrots can be added to the list of those seed dispersers, but many could have no place in the world until the big tropical fruits and nuts had evolved in the first place.

In fact many parrots very much resemble monkeys in their habits and manners. These are gregarious, mischievous and noisy. Many parrots, that is, not all of them. It will emerge in this chapter that these conspicuous parrots live in a specific type of environment, and even in a specific country (mainly). And their showy plumage has provided a wonderful tool for 'Mendelian' genetic studies.

Like Völker's 1937 paper, although on a much bigger stage, the work of Mendel was originally overlooked. Gregor Mendel was born in 1822 to peasant parents in a small, Czechoslovakian town. As a child he worked as a gardener, gaining horticultural experience, and eventually attended the Olmutz Philosophical Institute. In 1843 he entered an Augustinian monastery in Brno (Brunn, today in the Czech Republic), although he was later sent to the University of Vienna to study variance in plants. He returned to the monastery and his experimental garden, where between 1856 and 1863 he single-handedly cultivated and tested around 28,000 pea plants. Actually, he would have loved to work single-handedly, since his obligatory advisers were rather unimaginative and, although well-meaning, provided a scientific obstacle. From this monastery Mendel, nonetheless, in a paper of 1866, outlined most of the principal ideas on which the modern science of genetics was founded.

In short, Mendel initiated the theory of heredity and the idea of the gene. At first, these ideas were simply ignored. Then they were discredited when they were thought to counter Darwin's theory of evolution, which had appeared in 1859. In fact Mendel's ideas were not accepted until they were rediscovered at the end of the nineteenth century, and actually found to support Darwin's evolutionary theory. Indeed, the synthesis of Mendel's and Darwin's ideas became known as 'neo-Darwinism'. The structure of a gene, however, was not deciphered until the mid-twentieth century. Regardless, Mendel had discovered a form of 'practical genetics' that underpins livestock breeding, and can explain the plumage of offspring of different coloured budgerigar parents, for instance.

Returning to modern times, André Nemésio's interest in birds started when he was just four years old, when his uncle gave him a female canary. André was the eldest of four children from a poor family in Brazil and although his parents had no real interest in animals, his uncles did. Another uncle kept a large aviary with many bird species, and was a member of the local bird club. One year later that uncle presented André with a pair of budgerigars, and, following a clutch of eggs, his father considerately built two small 'flights' (long cages). André soon became introduced to the bird club, which was an unusual organisation because its membership stretched from bird hobbyists to conservationists in general. André's first recollection of this club was the homecoming of the first Brazilian to go to Antarctica in the early 1980s, with stories of whales, seals and krill. Hence André became acquainted with biologists of various fields. Of relevance to this chapter, the bird club was equipped with a good library for the time, and even at the age of nine or ten, André used to spend his afternoons reading the books there.

By the age of just twelve André Nemésio knew Mendelian genetics quite well, and used his understanding to direct the mating of guppies. His uncle's strong interest in guppies had rubbed off on him, and guppy breeding, like budgerigar breeding, is all about the genetics of colour. In fact André's knowledge of genetics at this age surpassed anything he learnt later at school, and was not improved upon until he studied biology at university, nearly ten years later.

In the meantime, at the age of fourteen, André returned to his budgerigars, and by integrating his knowledge of guppy genetics with ornithology he was able to contribute original ideas to bird breeding. His experiments began in a small room in his house – a bird-room conversion. André subsequently expanded his hobby and oversaw the construction of a more professional bird-room – a huge L-shaped brick building at ten metres by five, and with two floors, altogether divided into four rooms. In the two smallest rooms, André kept a mixed aviary. In the two elongated, larger rooms, he kept pairs of budgerigars in separate cages.

André's breeding experiments all went to plan. He predicted the colours of the offspring with accuracy. But it was important to keep a record of each bird's history since two yellow birds may carry different sets of genes, depending on the history of their parents. And concurrently, André would continue to trawl through the scientific literature and bird breeders' newsletters to discover parallel findings and inspiration for novel crosses. As I mentioned, he was just fourteen when he uncovered the Völker 1937 paper.

André was lucky not only to find a copy of this publication buried within his local university library, but also that the paper was written in German – he had studied elementary German at his school, and with the aid of a dictionary was equipped to translate the paper. This was certainly fortuitous. Unfortunately, the principal equipment used by Völker – an ultraviolet lamp – was beyond André's modest resources. Such a light was impossible to find in his part of the world, at that time at least. Fortunately, Völker's paper stuck in his mind.

It was not until 2001, at the age of thirty, that André fulfilled his ambition to recreate Völker's experiments – he managed to acquire an ultraviolet lamp. He waited until it was completely dark, entered his ever-operating large bird-room, and turned on the ultraviolet light. All was pitch black – the light was invisible to human eyes, although not to the eyes of the birds, of course (birds do see ultraviolet rays). André saw tens of live birds staring at him, or rather at his lamp, and they were *shining yellow*! There was no light visible to humans in the room, yet parts of the birds' plumage were plainly visible. This sight is fixed in André's mind still today. 'You can imagine my air of admiration,' commented André. 'It's a vision I'll never forget!' Note, 'Admiration', not 'Eureka!'

Not all the birds in André's bird-house were glowing yellow. And of the individuals that did glow, they did so from specific parts of their plumage only. So the problem relevant to this book is '*Why do some parrot feathers appear bright yellow to us when they are lit by ultraviolet light only?*' Remember, we do not see ultraviolet light – rays with wavelengths lying just beyond those of violet at the beginning of the spectrum – so the yellow feathers cannot be explained as ultraviolet reflection. But then not by yellow reflection either – there was no *yellow* light in the bird-house to reflect.

By this time, André was into his research career. As expected, he studied parrot biology and evolution. He had joined the Universidade Federal de Minas Gerais, which housed a 'reasonable' collection of healthy birds, as a PhD student. He began his doctorate project with a new and more thorough search of the recent scientific literature in his university library. On finding a paper on parrot biology, André would turn directly to the reference list at the end and tick all the cited papers that he did not have. Then he would try to get hold of them. He stressed the word 'try' to me – it can be difficult to obtain scientific literature in Brazil, and many searches were abandoned. But he did manage to obtain a new paper in the widely distributed, ornithological journal *Birds International*. The paper, by Walter Boles, the curator for birds at the Australian Museum in Sydney, was entitled 'Glowing parrots: need for a study of hidden colours'. André realised that Walter too had unearthed Völker's 1937 paper and, naturally, made contact with him. From that day, glowing parrots *did* receive study.

The night parrot – a key, negative result

Walter Boles was best known for his work on fossil birds, until, that is, his expedition to find the enigmatic night parrot. Unexpectedly, the story behind this did end with an ultraviolet lamp.

Since moving from the US to Sydney, Walter Boles has always been keen to keep the Australian Museum bird collection up to date through extensive field trips. I work mostly on marine invertebrates, so the concept of having all the species in one's specialised group described in the scientific literature is way beyond my grasp. But this is the luxury gifted to ornithologists, leaving their time free for studies of evolution, behaviour, ecology and physiology. Even so, his museum base had handed Walter a different priority – to keep an eye on endangered species.

Australia and New Zealand are countries where, in comparison, everything goes by contraries. In the case of parrots, the typically gaudy, gregarious cockatoos and lorikeets of Australia are in stark contrast to the characteristically quiet and camouflaged ground parrots of New Zealand. The New Zealand owl-parrot – the bizarre 'kakapo' – is indeed a lorikeet, although you would never know it after a visit from a flock of the more representative 'rainbow' lorikeets in Australia. A chorus of loud squawks and rustling branches accompanies the flashing colours as a flock hijacks a banksia tree. But the owl-parrot rather wears dingy brown and dark green, camouflaged plumage and works the nocturnal hours of an owl. It is a sombre, solitary species, lurking in the twilight or under the cover of the night, and digging burrows in the ground for its nest to stay completely out of sight. When the owl-parrot evolved, there were no dangerous enemies on the ground in New Zealand – no ground-based mammals. Hence the opportunity for a night prowler. Just as humans ruined that prospect in New Zealand with their pets that turned feral, in Australia the native animals – possums, dingoes and other marsupial hunters – had already closed most of the doors for ground-based parrots. All except the ground parrot and night parrot, that is.

Today the ground parrot is considered endangered, but is doing well in Tasmania. But for the night parrot the situation is far more serious. This bird has been considered extinct.

Australia has an unfortunate reputation for extinct species. That Australia is different from any other country in this respect is rather artificial. Extinct, Australian species tend to make the biological news, especially those to suffer the fate of European settlers. Because Australia has endured only 200 years of such catastrophe, notebooks were at the ready to record the details – science has been in full flow since the beginning of the Australian onslaught. But that explanation is still no excuse, and the demise of the Tasmanian tiger (thylacine) among other animals large enough to be noticed has become legendary.

When it comes to birds, Australian extinction legends have required some rewriting. Since Captain Cook's arrival, no mainland species of birds are known to have become extinct for certain. Recently, species designated as 'lost' have been 'found' – the noisy scrub-bird and Eyrean grasswren are two to re-emerge. But certainly, two uneasy cases remain.

The paradise parrot, with an extraordinarily colourful plumage, has not been scientifically documented since 1924, but every year we are fed tantalising stories of sightings from parts of its known, former range. And then there is the night parrot, even more enigmatic due to its nocturnal habits, although its natural history is the least understood of any Australian bird.

The night parrot was discovered in 1845, although this first scientific specimen was overlooked. Further specimens were caught in 1854, which were named by John Gould in 1861. In the 1870s a collection of sixteen night parrots from South Australia took place, but the last specimen to be obtained, and the only one in the twentieth century, came from Western Australia in 1912. A total of twenty-two night parrots are known to have existed through their corpses, although we are not sure where the remains of all these specimens are today.

Even worse for the night parrot, *sightings* from known collecting localities had also dried up by 1912, and the talk was of extinction. The Aboriginal opinion echoed this. Nonetheless, both the authoritative *Red data book of rare and endangered birds* and the *Threatened birds of Australia* theses give the night parrot a chance – their records read 'indeterminate'. What bigger incentive for Australian ornithologists?

Shane Parker of the South Australian Museum had led the most successful twentieth-century search party, apparently flushing out four birds believed to be night parrots. But like all other expeditions, this one returned empty-handed. The legend began to escalate, as did the potential rewards, culminating in a monetary prize. But the sightings reported by stockmen working at night were not considered substantial and that was that.

On 17 October 1990, two white Landcruisers driven by Walter Boles and lizard expert Ross Sadlier left the Australian Museum in the southern half of Australia and headed north-west, past Uluru (Ayer's Rock) in the very centre of the country and into Western Australia. Their passengers were Wayne Longmore, from the Queensland Museum, and Max Thompson, from Southwestern College in Kansas. Their aim was to sample the least biologically known parts of the country, to help fill the gaps in the museum's collections.

From where they hit the ocean at the coastal town of Broome, in the northern half of Australia, they headed through the Kimberly mountain ranges; again further north but now heading east again. Reaching the 'Top End' of Australia, now within the Northern Territory, Ross flew back to Sydney, leaving his reptile collection in the Landcruisers. Now it was all about birds.

Six weeks after leaving Sydney, Walter's expedition entered the return phase through western Queensland. He had chosen the scenic route, or rather the most scenic route. Highway 83 would divert them south from Mount Isa, a landmark still many hundred kilometres short of the Great Barrier Reef coastline. Just 100 kilometres into their journey south, the field team hit their brakes, distracted by a flock of

Australian pratincole (plover-like) birds. The birds flew and landed
down the road behind the Landcruisers, so Max turned one vehicle
around to follow them for a better view. Walter and Wayne stayed
where they were to minimise the disturbance to the birds. Max was
happy with his observations and brought his Landcruiser back to join
the other, parking just behind it. Walter got out of his vehicle and
walked back to speak to Max through the window of the passenger's
door. Then he turned away from the vehicle and just happened to look
down. There, on the roadside next to his Landcruiser, was the carcass
of a night parrot.

As a practising researcher myself, I am used to phone calls and visits
from science journalists, along with the question 'But what was your
reaction at the moment you made your discovery?' This is fishing for
the phrases, 'I could hardly breathe', 'My heart skipped a beat', or
something like that. Usually the bait is dangled until the researcher
'bites'. But against popular conception, few researchers have been
known to shout 'Eureka!' or the like. Personally, I am fond of 'That's
funny', but Walter, on this occasion, was further subdued. He calmly
passed the 'dead parrot' to Wayne through the window, and strolled
back to Max to report the find.

No one really showed much emotion, and normal collecting duties
resumed immediately, although the subject of night parrots did increasingly consume breaktime conversations. Meanwhile Walter was
preparing to make the most of the carcass while it was still in his
hands – it was found in Queensland and so its long-term future lay at
the Queensland Museum, in Brisbane. For now, though, it was on its
way to the Australian Museum in Sydney.

Very few parrots bear any resemblance to the night parrot, so the
identification stage of Walter's work, back in his office, was over in
minutes. In familiar terms, it looks like an over-sized budgerigar, but
they could never be confused. Perhaps its most serious contender is
the endangered ground parrot, but that lives near coastlines, not hundreds of kilometres inland, and has a number of different characters
to its body and plumage. Although this specimen was in poor condition – it was flattened from side-to-side and ants had taken much of
the muscle, although the plumage remained intact – its short claws

and tail were a giveaway that this *was* a night parrot. So what to do next?

There is debate over the relationship of the night parrot with the New Zealand owl-parrot, and so a sample of muscle tissue was removed for DNA examination to make a genetic comparison. Also, there were bones to be studied, especially the skull, although the disruptive work needed could be carried out only at the parrot's host institution, in Brisbane.

This serendipitous discovery of 1990 has taken us back to the year before Walter published his paper on glowing parrots and the need to study them. This is the time when Walter first began to experiment with a homemade light-box. The oversized chipboard shoebox, with a partially open front and black interior, contained a long 'black light' tube inside (a 'black light' is one with a high ultraviolet emission – which appears 'black' to us). This was Walter's semi-serious reaction to the Völker paper; equipment for an almost pre-pilot study.

Running out of things to do with the night parrot carcass, Walter casually placed it into the light-box, turned on the black light and . . . nothing happened. How exciting.

At this point you could be forgiven for thinking the last few pages, covering Walter's field trip, have been a complete red herring. But you would be wrong. Walter decided to place his light-box result for the night parrot into context, and from there a *positive* result emerged.

The bird section of the Australian Museum consists of three main rooms. The first, nearest the entrance to the annexe building, is filled with freezers and is where freshly collected birds are kept as they await taxidermy. The next room is large. Doors to various offices and a library interrupt the pale blue walls, but essentially the room contains rows of large, grey metal cabinets, filled with grey metal drawers. Pull out a drawer and a layer of stuffed birds is found, all with wings closed and identification labels at their feet. Walter opened the parrot cabinets, and species by species placed the specimens into the light-box. Many *did* glow. Walter made notes and even called in the photographic department to take photographs. The photographs spoke louder than any words that had been written on this subject. They were in fact quite deafening.

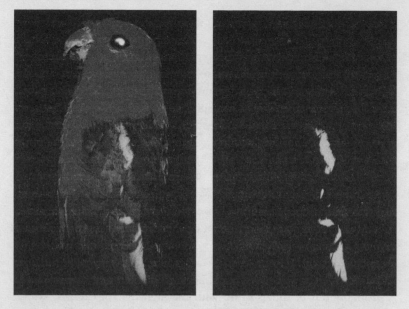

Figure 6.1   Photographs of the Australian king parrot made in the human visible range only, using white-light illumination (left) and using ultraviolet illumination only, as it appeared in Walter Boles's light-box (right). The glowing patch in the otherwise black ultraviolet photograph represents fluorescence.

Before leaving Walter's story, the last bird to be scrutinised with a black light lay in the third bird room. This room was very small, and was used to store miscellaneous bits and pieces – nests and eggs that had lost their labels, a stuffed albatross that could only be hung from the ceiling; that kind of thing. But it also housed a safe. The safe was a cast iron, classic 1920s 'bank-robber' type, and inside lay the CITES-listed specimens (CITES stands for the 'Washington' Convention on International Trade in Endangered Species of Wild Fauna and Flora, which aims to protect certain plants and animals by regulating and monitoring their international trade to prevent it reaching unsustainable levels). This was obviously the museum's endangered bird collection, and another reason why bird collections differ from those of marine invertebrates – they are the targets of unscrupulous collectors. The curator of the bird collection at the Oxford University museum explained to me only recently how they once foiled an attempt to steal their (extinct) great auk egg. So Walter was taking no chances with his scientifically and commercially valuable specimens.

Walter did obtain a positive result from one bird in the safe – the golden conure. This completes a rather ironic story. Walter was examining a Brazilian bird from his Australian office and André was working on the Australian budgerigar from his bird-room in Brazil. Nevertheless, this combination became a creative one.

'Over to you, André'

Since time became restrictive for Walter, who was committed to and consumed by several other ornithological projects, including management of the Australian Museum's extensive and scientifically important bird collection, André took it upon himself to make detailed, biological sense of Walter's findings. In 2001 the two corresponded extensively by email, and Walter was happy to hand over his photographs.

André had taken on the role of editor of the Brazilian-based *International Journal of Ornithology* (formerly known as *Melopsittacus*). He had thought about publishing his thoughts on parrot colours, which he knew would be most useful to Mendelian-type studies. Now the time was right, as such an article would also provide the perfect place to reveal the results of his ultra-violet lamp and Walter's black-light-box. There was a story of evolution here, too.

Walter's photographs revealed parrots as we know them, with bright patches of reds, yellows, greens and blues, but also as we have not seen them before. André arranged Walter's pictures into two columns – ordinary, white-light photographs on the left; ultraviolet-lit photographs on the right. The ultraviolet-lit photographs appeared either as completely black, where the glow was absent, or as black with an irregular-shaped patch of bright yellow. To discern the rest of the bird, the normal, colour photo had to be observed. In no case did a parrot glow throughout its entire body, and soon it became clear that the glowing patches only ever appeared yellow under white light.

Walter had revealed that the night parrot did not glow. This provided the inspiration to test other nocturnal parrots such as the ground parrot, also from Australia, and the owl-parrot (kakapo), kaka and kea from New Zealand. None produced a glow. Although some of these species did possess yellow pigment in their feathers, they would never glow under an ultraviolet light. That result was interesting because ultraviolet light is virtually absent at night, meaning that if the feathers of these birds were capable of glowing, they would never do so in the wild. Was this evidence that the glow played a role in the bird's behaviour? Even more enticing from this respect was the case of the golden conure.

The golden conure, also known as the Queen of Bavaria's conure, is a medium-sized and average-shaped parrot, active during the day. Being CITES listed, the golden conure is obviously rare, confined to a small range in north-east Brazil as a victim of deforestation (although reports suggest they are on the move, searching for new habitat). Regarding its colour, other than the green ends to its wings the golden conure is entirely yellow. *Uniformly* yellow over its whole plumage, under white light. The ultraviolet photograph, on the other hand, revealed something odd. Where all other parrots that fluoresced always showed an *entire* patch of one colour (yellow) glowing, the golden conure glowed in only *part* of its yellow 'patch'. This meant that the glow was not simply a consequence of a yellow pigment. For the first time it was revealed that there are both yellow pigments, and a quite separate glowing factor.

Even more interesting was the position of that glowing part of the golden conure, at the nape (the back of the neck). This is generally a part of the bird important for mating – colours employed here, along with other parts of the head, are used for courtship among parrots. Golden conures are particularly social among parrots, with pairs existing in almost constant body contact and usually found preening each other. They tend to inhabit the very tops of trees, where the ultraviolet content of sunlight is at its highest (further into the canopy it becomes scattered by leaves, being more susceptible to scattering than other wavelengths). In fact the evidence from the golden conure all points towards a *courtship* function for the yellow glow.

Considering the cause of the parrot's glow

Between them, André and Walter had amassed enough evidence to suggest that the parrots' glow was beyond the biologists' feared 'I' word – incidental. Nocturnal parrots, deprived of the ultraviolet fuel for this effect anyway, did not glow. A day-active parrot, in receipt of the energy for the visual effect, did glow, and in the communicative parts of its plumage at that. It began to seem as if the glow was not simply a by-product of a yellow pigment, but had evolved independently, in its own right. The tables were turning. For the first time the parrot's glow was being considered as part of its behavioural repertoire. And that's justification to examine the *cause* of the yellow glow, and to attempt to answer the problem central to this chapter. Now, it would appear, any work on the parrot's glow would not be a waste of valuable, scientific time.

Like mammals and reptiles, no bird can bioluminesce. And since the bright yellow patches occur even on *dead* specimens lying in museum collections, we can be confident that no bioluminescent reactions are taking place (bioluminescence results from chemicals produced by living cells, unless the stored chemicals in dead specimens are deliberately mixed together). Pity, because the *generation* of light characteristic of bioluminescence could explain a yellow appearance without a yellow light source. Some mushrooms glow yellow through bioluminescence even in complete darkness. So now our other options for colour production should be explored.

Bird skin is thin, elastic and loosely attached to the body, allowing birds the freedom of movement needed for flight. The skin, of course, tends to be covered mostly by feathers, although bald patches can be coloured by pigments or structural hues. But feathers are the usual source of bird coloration. The evolution of feathers in birds has been accompanied by the development of complex systems for producing colours and patterns. Feathers are made of the protein 'beta-keratin', as found also in reptile scales, which is a strong and stiff yet lightweight and flexible building material. These characteristics are ideal for flight, but also the air-filled, spongy lattices that fill the insides of feathers provide excellent insulation, and the finest branches of a feather maintain the surface tension of water drops and so provide waterproofing (the large drops simply roll off). But in addition to its structural properties the feather may play a visual role via pigments of two main types – carotenoids and melanins.

Carotenoid pigments provide reds, oranges and yellows and enter feathers from a bird's stomach (sometimes with chemical processing *en route*) – they are obtained only from diet. Flamingos extract their carotenoids from ingested blue-green algae. However, carotenoids do not reflect light – they only absorb it. They absorb blues and greens allowing reds and yellows to pass by, which are reflected back into the environment by white 'scattering' structures (the spongy lattices) deep within the feathers. Darwin suggested that bright feather patterns in birds are the result of sexual selection through female preference for extreme characters, including plumage colour. More recent models assert that such ornamental characters function in sexual selection as *honest* signals of individual quality – bright colours indicate a fit and healthy individual, with 'desirable' genes. Not surprisingly females of carotenoid-coloured birds choose males with the brightest colours because this indicates the most efficient food-finders and parasite-avoiders.

The other main type of pigment found in birds is melanin. As described in the Ultraviolet chapter, melanins are responsible for black and brown colours. They are not derived from diet and so must be synthesised by the birds themselves, but still act as reliable indicators of mate quality. In contrast to carotenoids, melanins indicate the social status of a male within a group of birds, rather than body condition. Male sparrows that take part in numerous aggressive encounters during moulting, for instance, develop larger black melanin-based throat 'badges'. These are the strongest males, and so potentially the best fathers in the eyes of a female. Birds moult, and not surprisingly these mating signals appear most prevalent during the breeding season. But returning to structure, and remaining with colour, evolutionary plasticity has resulted in the alteration of the *architecture* of the feather.

Feathers consist of a shaft or quill and a plume – the barbs and barbules. The plume is responsible for the brilliant iridescence of hummingbirds, where it is loaded with multilayer reflectors and dif-fraction gratings, and even the ultraviolet reflection from a black lorry parrot. This parrot appears only black to us, and the presence of black 'melanin' pigments can explain why. Except these pigments are dis-tributed within a spongy matrix of keratin and air that packs the inside of the feather plume. In any direction, the arrangement of keratin and air will form a series of layers, with dimensions just right to be a mul-tilayer reflector for ultraviolet light. Fellow birds observe the ultraviolet reflections – all is not as it first seems in the black lorry.

The Danish ornithologist Jan Dyck reported the concept that such a spongy structure should approximate a multilayer reflector in many directions in 1971, although the great Indian scientist C.V. Raman had suggested the idea in 1935. Prior to this, the spongy structure respon-sible for many blues in feathers was considered a random set of holes and, consequently, a medium for Tyndall scattering (the haphazard, unordered reflection of light rays from randomly arranged elements smaller than the wavelength of light, where shorter wavelength blue rays are reflected more often than the longer wavelength reds). But recently a team of biologists and mathematicians, including Rick Prum from the University of Kansas, began a theoretical treatment that veri-fied the presence of multilayer-type reflections. So the rays leaving the feathers constructively interfere, rather than avoiding interactions as in the case of Tyndall scattering (although the absence of a metallic appearance remains slightly puzzling).

Structures causing blues are often combined with yellow pigments in bird feathers to provide greens in the manner of the green tree frog. It would seem that the evolution of green pigments is problematic also in birds, but the blue and yellow combination is common – it occurs in the green feathers of budgerigars, for instance. Different, independent genes control each element (blue and yellow) of the green colour. And this is why budgerigars are model species for Mendelian genetics experi-ments – they can be bred so that the gene for the yellow pigment or blue structure is not expressed, with the additional possibilities for the blue reflecting structure to increase in size to cause 'white' scattering, and the expression of varying proportions of melanin pigment to add an element of tone.

With this knowledge of bird pigments and structural colours in mind, we should employ NanoCam on the golden conure to try to identify one of these colour possibilities, or even to search for what is perhaps unknown, since there is plenty in this case that simply does not add up. Does the bright yellow patch on the nape of the golden conure contain pigment-filled or structurally coloured feathers? Or is there something else not covered so far in this book?

Detail of the colour factory, and the solution

Weaving through the canopy leaves, the optical probe of NanoCam approaches a golden conure from behind. The feathers of the yellow tail are first encountered, and out of curiosity, NanoCam passes between the keratin molecules of a barbule's surface and encounters molecules being worked by the sun's rays. As a beam of sunlight, containing a complete range of wavelengths, is imaged on the monitor, it promptly strikes one of the molecules. Electrons in the outer shells of some atoms are catapulted into new, further outer shells as the energy is grabbed from the violet, blue, green, orange and red rays in the beam. These colours are all absorbed – removed from the beam of sunlight. As the electrons drop back down to their original outer shell, the energy taken from those colours converts to heat and NanoCam begins to warm. The ultraviolet and yellow rays in the beam, on the other hand, were not removed in this way, and were rather reflected from the molecule – reflected in all directions. What we have observed are ordinary yellow pigment molecules that are responsible for the yellow appearance of the feathers in this region of the plumage. Beyond our detection, these yellow feathers are also reflecting ultraviolet light, and that is very much part of the bird's plumage coloration too.

NanoCam pulls out and travels along the parrot's back until it reaches similar-looking feathers in the region of the nape – the area that glowed under a black light. Again the camera optic winds its way between the outer surface molecules of a barbule and enters the interior.

The interior of this barbule appears different to that of the back feather examined. Here there are more molecules being influenced by light. The interior of this barbule is packed full of atoms, with electrons jumping around all over the place.

The New Zealand scientist Ernest Rutherford is to the atom what Darwin is to evolution and Faraday to electricity. He was first to discover the structure of the atom. He also found that an atom consisted mostly of . . . nothing. There was a central nucleus and electrons that orbited, but between there was nothing – lots of nothing. Rutherford likened the electrons of an atom to a few flies buzzing around within a cathedral, with those of the outer shell flying close to the walls. This gives an appreciation of scale for our feather investigation. Within this feather barbule there are many electrons leaping beyond the walls of their 'cathedrals', only to return back within the walls swiftly.

With some focusing, NanoCam singles out a yellow pigment molecule as it occurred in the tail feather. Soon, many of those pigment molecules are detected. But interspersed with these are other atoms affected by light. These belong to another, very complex molecule (since everything is tightly packed within this barbule, it is difficult to tell whether or not this molecule is attached to a yellow pigment molecule). NanoCam focuses on one of those atoms.

A beam of sunlight strikes the barbule and passes, unaltered, through the outer surface of keratin, which acts like a window. All wavelengths hurtle towards the atom in view and crash into its orbiting electrons. This time, most wavelengths are not affected in any way, and continue along their path through the barbule (but may later encounter a yellow pigment). Some wavelengths in the sunlight, on the other hand, are very much affected. These lie in the ultraviolet region of the sun's spectrum.

NanoCam views the electrons of the outer shell as they are again propelled into new, outer orbits after seizing the energy of the ultraviolet rays. They put an end to the ultraviolet light – there will be no ultraviolet coloration in this region. That's interesting because from the other yellow region – the parrot's tail – ultraviolet light *was* reflected, making another visual contrast between the two regions in addition to the glow (to other parrots, but not to us).

Just as for the yellow pigments, the energy needed to drive an electron from its outer shell in the first place becomes converted to heat when the electron drops back down (they fall back down within nanoseconds, because the atom with the *new* electron arrangement is extremely unstable). Being so close to this atom, NanoCam does begin to warm. But the energy contained within an ultraviolet ray does not *all* become heat.

The atom in view is held firmly in place – the molecule to which it belongs has a particularly rigid structure. It cannot move around much. As it attempts to release the energy absorbed from an ultraviolet ray, it moves only slightly – less than the movement observed in the unrestrained yellow pigment atoms. It is this movement that generates heat. NanoCam further scrutinises this atom as it struggles to vibrate sufficiently to dump all of its energy gained as heat. It moves only enough for *some* of the energy to be converted to heat.

So what happened to the remainder of the energy absorbed from the ultraviolet ray? Since energy cannot be destroyed, where did it go?

Remember that the short ultraviolet rays contain the most energy in the spectrum, and the longer red rays the least. Remove half the energy from an ultraviolet ray and you have a red ray. More specifically, the atom of interest here could move just enough to remove a third of the ultraviolet ray's energy and convert it to heat. And guess what has two-thirds of an ultraviolet ray's energy? A yellow ray, of course. The

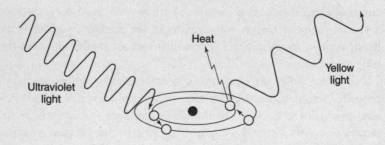

Figure 6.2   A diagrammatic atom in the process of fluorescence. An electron (white circle) of the atom's outer shell is promoted to a new outer shell but falls back to its original shell in 10^{-8} seconds.

energy remaining was dumped as yellow light. NanoCam had observed an ultraviolet ray strike the atom, a yellow ray leave the atom, and the emission of some heat. This process is known as *fluorescence*.

Fortunately the yellow *pigments* did not absorb the ultraviolet rays, which became available to fuel the fluorescent atoms. Accordingly the combined effect of the barbule's molecules in the nape region was to directly reflect yellow rays, to convert ultraviolet rays into further yellow rays, which also emanate from the feathers, and to absorb all other wavelengths. This, then, explains how a parrot can appear yellow to us when lit only by ultraviolet light – we do not see the ultraviolet but we see the yellow where ultraviolet rays are converted. We have solved our problem for this chapter – *fluorescence* is the answer.

To give due credit, Völker was aware of fluorescence and recognised it as the cause of the parrot's glow. Fluorescence was first identified in England many years before, when the mineral *fluorite* was exposed to ultraviolet light and glowed blue. Hence the name 'fluorescence'. By 1937 fluorescence was known in many minerals, and had even become a common means of mineral identification. Probably Völker came to observe fluorescence in his parrot collection at that time after borrowing a black light from a mineralogist working nearby in Heidelberg. Still, there is a little more to consider about fluorescence than just its occurrence in nature, in particular what it means to biology. Mineral fluorescence, of course, must be incidental – a glow can have no purpose for a rock – but André and Walter had uncovered tantalising evidence that this may not be the case in parrots. And if fluorescence *did* have a part to play in parrot behaviour, then it would be well worth tracing the evolution of this visual effect.

Back to the story – the evolutionary tale

Now we can consider the story of evolution that had sprung from Walter's photographs of parrot fluorescence, André's knowledge of parrot breeding and colours, and a little geological history.

Parrots are divided into Australasian, African and South American groups. Two hundred and fifty million years ago, all the continents were fused into a single landmass, which later divided into two great continents – Gondwanaland and Laurasia – separated at the Earth's equator region by the Tethys Sea. Gondwanaland consisted mainly of Australia/New Guinea/New Zealand, India, Antarctica, Africa and South America. About 120 million years ago, the Australasian plate and Antarctica broke away entirely and the Indian Ocean was formed. This event was just preceded by the evolution of birds. The Atlantic Ocean opened up completely between South America and Africa about 80 million years ago. But the real events that affected birds came later.

Birds went through a massive bottleneck 65 million years ago at the K-T boundary, the period famed for the dinosaurs' demise and the rise of mammals (supposedly triggered by a cataclysmic meteor impact), where the *Archaeopteryx*-type birds also became extinct. The ancestors of many of the modern groups of birds, on the other hand, survived. Ancient among modern birds, the parrots are one such group to evolve long before, and to survive, the K-T boundary – probably they evolved in Gondwanaland (this was also the case with penguins and the ostrich/emu/rhea group, for instance).

Walter had revealed, among his fluorescent group, species from both Australia and South America, symptomatic that the gene for fluorescence is rather ancestral in parrots. André took this idea a stage further, through consideration of all forms of colour in these birds, and suggested that the rule in parrots is to lose existing colour characters such as fluorescence, and not to acquire new ones. He saw the ancestor of all parrots as a bird containing every means of producing the colours found in them today. One alternative view is that a gene for fluorescence evolved twice, independently – once in an ancestor to South American species, and once in an Australian species, somewhere between 120 and 80 million years ago. André, nonetheless, makes a good case for *his* hypothesis.

Walter's photograph also revealed that fluorescence is common in Australian species active during the day, but rare in South American and African parrots (although parrots are not well represented in Africa). For instance, African lovebirds and all macaws (of South America) do not fluoresce, yet these *are* brightly coloured. It also appeared that any parrot without yellow colours did not fluoresce – the genes for yellow pigment and fluorescence seemed to be associated but, as we observed in the golden conure, not mutually exclusive.

This consideration also pointed André to another trend that he had previously overlooked – that South American parrots tended to be mainly green, and Australian parrots tended to show little green. Maybe the Australian species had largely evolved along a path to conspicuousness and the South American parrots (with some obvious exceptions) towards camouflage. Like the golden conure, brightly coloured parrots tend to occupy the top of a forest canopy, where light is rich in all wavelengths or colours. Green parrots usually live in the lower regions of a canopy or in the understory, and here the ambient light is also mostly green since it is the colour rejected by leaves (ultraviolet is the most reduced colour in these zones). This certainly helps to explain the occurrence of fluorescence, which arises only in places where ultraviolet light is strong. And noise may further fuel this argument. The golden conure is among the most vocal of South American parrots, and the brightly pigmented macaws can compete for volume too. Most Australian parrots are brightly coloured, and I can vouch that these are loud. This is all good news for the functional nature of fluorescence in parrots. The evidence does point to a means of enhancing conspicuousness. After all, fluorescence could potentially push the emanance of yellow beyond the 100 per cent level (i.e. the portion of yellow in the incident light is surpassed) – as colour-efficient as animals can ever be.

So the premise to be tested is that the ancestor of all parrots evolved a fluorescent pigment for use in courtship, but where some subsequent evolutionary lines headed towards inconspicuousness, they lost their fluorescent pigments. From this point fluorescence did not re-evolve, being a real one-in-a-million chance event (unless it re-evolved once in the more derived South American group). This would suggest that parrots first evolved to occupy a vacant 'way of life' on Earth where conspicuousness was key; where sexual selection rose above natural selection. Originally their population increases through efficient courtship and breeding would have more than offset their losses from predation, thanks largely to a means of predator avoidance – flight – as opposed to camouflage.

The fluorescent pigment has been examined by biochemists, but the report so far is that the structure of the pigment is very unique and requires considerable energy to make. Behavioural experiments on budgerigar fluorescence have also begun, although up till now these have produced conflicting results. Since this work is in the fine-tuning stage, the best suggestion is perhaps to 'watch this space'. Hopefully we will know whether fluorescence has a function in parrots soon.

It would be interesting also to examine the *other* effect of fluorescence on a parrot's behaviour – fluorescence not only increases the reflection of yellow, but additionally removes ultraviolet from a feather's spectrum. This could provide colour and brightness contrast if the surrounding regions of a fluorescent patch do reflect ultraviolet light, and so further increase the visual vocabulary of parrots. This does, inconveniently, complicate matters for the biologist conducting behavioural tests. Namely, if parrots react differently when ultraviolet light is removed from their environment, is this due to the absence of ultraviolet reflection or to fluorescence?

Finally for parrots, genetic studies are resolving, in ever-increasing detail, our understanding of their evolutionary trees. It is beginning to appear that the Australian parrots branched off from the base of the tree, as expected, that the South American species branched from the top of the tree, being the most derived, and that the few African species branched from in between. But interesting is the position of the golden conure – it lies at the very top of the evolutionary tree, the most derived even among South American parrots. This could suggest that the fluorescent pigment evolved twice, independently – once in the Australian group, from where it also became lost (probably on more than one occasion), and again in the golden conure and perhaps its closest relatives. Still, André Nemésio's theory at least fits for the majority of cases. Certainly more parrots should be placed under an ultraviolet light in the future. As you see, although all species of birds on Earth may have been described, there is still plenty of other ornithological work remaining.

Hornbills

Earlier I mentioned that most parrots closely resemble monkeys in their habits and manners. Both evolved to fully exploit the fruits and nuts of tropical tree canopies. Moving on through geological time, it is well known that the great apes evolved from some of these monkeys, where forest-dwelling fruit eaters descended to the ground, moved out to live in open savannah, adopted a carnivorous diet and developed advanced forms of social organisation. Although the parrots never mirrored *this* evolutionary development, another group of birds have – the hornbills.

Hornbills are large birds of Asia and Africa, reaching up to nearly two metres in height, and characterised by a bulky extension of the huge upper bill. Most hornbills have black, white or grey plumage but strikingly colourful bills, although, contrary to this description, they are not related to toucans. Their bones have much larger air-spaces than those of most birds, and, although appearing awkward while flying, they do manage to fly long distances. Flight is accompanied by a rush of air through their characteristically open-based wing quills, which is distinctly noisy (comparable to an aircraft if the bird is nearby). The food source of most hornbills is berries and fruit, although some will also eat meat. Maybe the success of the smaller hornbills in the treetops of Africa can explain the failure of African parrots to diversify there. Maybe they compete for the same way of life.

Although most hornbills live in trees, uniquely in the African savannah they have also moved to the ground. The southern ground hornbill lives in cooperative groups, and has turned to a carnivorous diet, feeding on insects, snails, spiders, snakes and birds and mammals up to the size of hare. It is territorial, proclaiming possession of its range with deep booming calls. Sunbathing and dustbathing are favourite pastimes, and as a bird flops to the ground with wings outspread, other hornbills will walk around and preen them, like hairdressers. Heaviest preening, though, is reserved for their courtship rituals.

North of the equator, in northern Africa, the southern ground hornbill has been replaced by, believe it or not, the northern ground hornbill. These appear less socially organised than their southern relatives, but continue the preening trend. And in this species an unusual, visual trait has evolved – they actively *apply* some of their colour to their bodies themselves, in the manner of make-up. This is unusual for animal colours in general.

Figure 6.3 The great hornbill.

At the base of the northern ground hornbill's back lies a preen gland – a bare bump, surrounded by a tuft of feathers, that secretes cosmetic oils. During the mating season, these birds twist their heads around and gather the yellow-pigmented oil on their bills. Then the bill is wiped over the white wing feathers, which become stained yellow, although the bill receives colour in the process too. And the more yellow the wing feathers, the more attractive the individual to a potential mate.

The yellow, cosmetic oils are also fluorescent. They glow yellow under an ultraviolet light, again enhancing the yellow pigments signal through conversion of the ultraviolet rays in sunlight. Since yellow is obviously a sexual cue, one can only assume that the extra yellow provided by fluorescence serves to increase the attractiveness of the individual.

This application of 'make-up' is shared with the large fruit-eating hornbills of Asian rainforests – in fact there may be an evolutionary relationship here, too. The classic example is the great hornbill, the representative species most commonly exhibited in zoos. This species, found in India, also applies fluorescent make-up when it wants to look at its best. And for this species there is an extraordinary parallel in the art world.

In pre-twentieth-century Indian art, the paintbox contained a variety of media loosely but generally termed 'gouache'. The British Museum in London holds one particular painting in gouache from the early seventeenth century by the accomplished Indian artist Manṣūr, famed for the accuracy of his animal studies. The painting is of a great hornbill, standing on the ground, and involves a number of different pigments, more than could be found in the European paintbox of that time. A close inspection of some pigments in this gouache, namely the green, red and yellow shades, will reveal why.

Two shades of green exist in the background flora. The pale greens of the grasses in the foreground derive from the mineral malachite, and in particular its copper component – all fairly normal. The dark green, on the other hand, is not what it seems. Here we return once more to the green tree frog scenario, where for want of a naturally available dark green pigment, deep blue and yellow pigments were mixed together. Under a microscope, individual grains of these two colours are evident, and like the scales of butterflies, adjacent grains are combined into a single pixel in the naked eye in pointillistic fashion, in this case to make dark green (the *pale* blue of the tree frog combined with yellow to produce a paler green). The deep blue pigment is indigo, an organic chemical extracted from the fermented leaves of the indigo plant, actually akin to the blue woad used by the ancient Britons. The yellow pigment is orpiment, or arsenic trisulphide, which, although similar to the orpiment employed by the ancient Egyptians, did not really take off in Europe, possibly because of its lethal effect on all things living.

The minimal crimson pigment on the bill became known as lack. Lack was extracted from the 'lac' insect, which tragically owes its name to its pigments, or rather their extraction. 'Lac' was derived from the Sanskrit *lakh*, meaning 100,000 – vast numbers of insects were needed to produce the dye.

The yellow on the wing feathers and bill of Manṣūr's great hornbill represents this bird's make-up. It is a different shade of yellow to the orpiment used as part of the dark green, and appears to match closely the hue of the hornbill's actual preen oil. This pigment is called peori, or Indian yellow, and by 1890 it was outlawed for reasons of cruelty.

Indian yellow derives from mango leaves, although it also involves cows. Cows in the village of Mirzapur, in the north-east Indian province of Bihar, were fed mango leaves and the pigment was extracted from their urine. Not pleasant for the pigment manufacturer, but even less so for the cow – mango leaves are not nutritional for cows, which were fed nothing else for fear of affecting the pigment, and so proceeded to suffer diet deficiencies. Not surprisingly this practice was considered cruel and banned in the more humane age. Remarkably, Indian yellow is *fluorescent* – the only fluorescent pigment in gouache.

The artist Manṣūr had unknowingly reproduced the great hornbill with scientific accuracy. Certainly he could not have known of the fluorescent properties of either the hornbill's feathers or his Indian yellow pigment (nor would he have known of fluorescence) since ultraviolet-only lamps did not exist, of course (ultraviolet was a concept of the future). He may, just, have observed a unique glow about his yellow pigment. But if his painting is placed in the dark and lit with a black light today, regions of the bill and wing feathers glow, just like they do on the hornbill itself, and as they did on the budgerigars in André Nemésio's original bird-house.

Ghostly scorpions and Frankenstein fish

Fluorescence is not confined to parrots and hornbills among animals. Indeed, it is not confined to birds (which do exhibit a fluorescent diversity). This jogs my memory of a case of fluorescence I knew very well even before joining the parrot conversations of Walter and André – the case of the scorpions.

Scorpions are animals almost synonymous with fluorescence. In sunlight, fluorescence may impart a greenish tint to a scorpion's colour, but under a black light the animal glows strongly over its entire body, appearing like a ghostly figure. Not surprisingly, where scorpions are kept as pets, a black light can be found customarily in their enclosure. The fluorescence derives from an extremely thin layer in the scorpion's exoskeleton, which develops after moulting. As a new exoskeleton hardens, its fluorescent quality increases. This indicates that the fluorescent molecules are either secreted purposely by the scorpion shortly after moulting, or are a by-product of the tanning process that hardens the exoskeleton.

Although the jury is still out on whether the scorpion's fluorescence has a function, it would appear unlikely, since scorpions are without image-forming eyes and are basically nocturnal anyway. They do, nonetheless, possess light sensors or 'ocelli' that have excellent low-light sensitivity, and some researchers suggest that the levels of fluorescence *are* detected and provide an indication of ultraviolet light levels in the atmosphere. During high levels the scorpion could go into hiding, since ultraviolet light can be damaging to biological tissue. All very well, except that it is the UV-B form of ultraviolet that is most damaging in this respect, and scorpion fluorescence is stimulated by UV-A – different wavelengths. Whatever the case, scorpion fluorescence is extremely useful when collecting scorpions in the field – powerful ultraviolet lamps are used to illuminate the ground, and any scorpion in their path will appear totally maladapted to vision, glowing brightly in the dark. But then arachnologists with ultraviolet torches were never a selection pressure during scorpion evolution.

Scorpions had always been model animals when it came to fluorescence, until the beginning of the twenty-first century. Now we are taking the subject more seriously and are finding fluorescence in a whole range of animals from giant clams and shrimps to corals and butterflies, possibly spurred on by the discovery of 'GFP'. 'Green Fluorescent Protein' is a gene from a jellyfish that codes for a fluorescent pigment, which glows green. This gene–pigment relationship is unusual, nonetheless, and has taken fluorescence to new heights of scientific interest – probably it is the most studied subject on nature's colour palette today.

Ordinary pigments begin their life history as a sequence of aminoacid bases on a strand of DNA – a gene – that becomes translated to a free protein molecule. That molecule must then undergo reactions to gain further groups of atoms, sometimes including metal ions. But GFP is different. In this case, the molecule is translated *directly* from its gene. No further reactions are needed. So how does this characteristic make GFP useful?

Francis Crick famously commented, 'DNA makes RNA makes protein.' The work of DNA is to make proteins, and this is achieved with the aid of a smaller and simpler nucleic acid known as RNA. The DNA double helix separates or unzips along part of its length, and one of the separated chains makes a compliment to itself – RNA. Then the RNA, whose structure reflects that of the DNA, ensures the joining of amino acids in the appropriate order to form a certain protein – that coded by the gene. This protein may go on to become a pigment, through a series of chemical reactions, or, in the case of GFP, it will be fluorescent *just as it is* (with the acquisition of oxygen). Well, it does reshuffle its atoms within itself to some extent, altering its molecular shape, but this activity is self-contained. So GFP becomes interesting to geneticists studying gene expression – if you want to know whether a gene is being expressed, why not use GFP, where the presence or absence of a glow will provide a quick 'yes' or 'no' answer?

Not surprisingly, Green Fluorescent Protein has been a hit in the subject of genetics. And like the orchid growers who implanted the gene for bioluminescence into their plants, some scientists are carrying out the much simpler task of inserting the GFP gene into a variety of animal embryos. I remember a front-page article from a national (broadsheet) British newspaper in 2002, which led with the title 'Frankenstein fish will glow in the bowl'. The article included a picture of four small, minnow-like zebra-fish, glowing green against a black background. Of course the fish had expressed their GFP genes – they had manufactured the protein for which it codes – and were fluorescing under ultraviolet light. It was an eerie, unnatural sight that sparked concerns from ecologists about the genetic contamination that may result should these engineered fish enter natural waterways and breed with related, native fishes (the GM salmon argument). Fortunately GFP is also employed towards more constructive means, which may ultimately lead to improved human health.

And finally, if one requires a quick demonstration of fluorescence, just look to highlighter pens. The reason their ink is so bright is that the effect of its normal pigment is amplified by a fluorescent emission in the same colour. But fluorescence should not be confused with phosphorescence, possibly more prevalent in commercial products, as seen on the glowing hands of a watch at night, or glow-in-the-dark stickers. This is a slightly different process, again involving the absorption of ultraviolet light, but this time the excited atom is more stable – an outer electron can remain in a new outer shell for quite a while – so that the time until the energy is released is much longer. The result is a glow after the light source has been removed. Phosphorescence is not known in nature, although the term has been used mistakenly to describe marine bioluminescence on more than the odd occasion. Bioluminescence remains the natural option for that glow-in-the-dark look.

A tonic for Darwin

This chapter suggests that an eye may become attuned to seeing certain wavelengths particularly well. If birds employ fluorescence for courtship, they may have evolved a private signalling system that leaves their predators out of the loop. Through pinning much of their signalling hopes on their yellow patches, they may have evolved a visual system that is highly sensitive to yellow. And since they impart more yellow light than even exists in sunlight, they can communicate in fine detail through this colour. But their bird predators are generalists and so cannot afford visual sensitivity for a particular colour, because they must remain on the lookout for the entire spectrum. So a predator can be blind to a subtle but meaningful change in a bird's yellow signal – its secret code. As a way to increase its vocabulary, a bird can avoid using the entire spectrum by boosting its display of a single colour, and all quite safely. The golden conure's predators would benefit from extra sensitivity to yellow.

We suffer the same shortcomings of the predator in this sense. We look at the back of a golden conure and see no difference in its colour emission. The golden conure, meanwhile, probably reads a message broadcast from the feathers of its nape. This is an excellent demonstration that no eye can be universally perfect – every eye must have its limitations, which ultimately become its Achilles heel. Limitations become the target of evolution-selection pressures to other animals in the community. In theory, a visual system could be designed to be limitless, but no animal could afford to house the necessarily gargantuan brain. Compromises are necessary. Once more, *the generic eye is not perfect* – the perfect eye is just not practical. Darwin was wrong to think it could throw a spanner in his evolutionary works in this respect. If he had known the finer details of nature's colour arsenal, he would have realised how they exploit the *weaknesses* of eyes, and would have slept more soundly.

When it comes to deceiving the eye, the next chapter will consider another snake-in-the-grass, only this time quite literally. A case of mimicry will be scrutinised with the ultimate trick up its sleeve. The eye will be fooled into seeing a colour that actually is not really there – probably the most remarkable case of colour in animals, and one hot off the scientific press.

COLOUR 6

orange

The problem:
How can an orange, black and white banded snake appear green?

Milk snakes are one-metre-long reptiles from the New World – North, Central and South America – that are active during the day, live often against a background of grasses, and do not conceal themselves like other snakes. They are preyed upon by small, carnivorous mammals, birds of prey and even other snakes, and are known to be among the fastest snakes in existence. In terms of appearance, milk snakes are distinctive only for their colours, particularly their vivid orange (or in some species red) bands. They appear black, white and orange striped throughout their body lengths. When Adrian Thomas, a colleague in Zoology at Oxford, casually mentioned that he thought he saw green when he observed a milk snake moving, I thought, 'That's odd,' although at the time I did not realise why. Fortunately I decided to follow up this anecdote with a series of experiments (apologies from the beginning, but this chapter will weigh in heavily with one of my *own* scientific adventures). And surely the first stop for any investigation of this type has to be colour mixing.

Figure 7.1   A Honduran milk snake.

Another variation on colour mixing – movement

The blue and yellow macaw, a giant South American parrot, has a blue back (including the back of the head), wings and tail, while its chest and legs are yellow. Its face is black and white, but we will overlook that here. Viewing the macaw from its front while it sits on its perch, the brain interprets the shapes formed in blue and yellow.

The yellow shape is a rugby ball with legs. The blue shape is an elongated 'V' that includes the tail. The point is, neither the yellow nor the blue shape say 'macaw' in the brain's dictionary of shapes. To achieve the stereotypical shape of a parrot, the brain must combine the yellow and blue shapes and separate them from the background. Now we can consider something else – what happens if two colour patches in an environment can affect the *same* pixel in the brain? In this case, the colours of each patch would be combined.

Many animals appear a different colour from that of their reflected rays. The Atlas moth of South East Asia, the world's largest moth, appears with regions of mustard and grey. A closer inspection, however, reveals neither mustard nor grey pigment. In pointillistic fashion, the mustard region is packed with only brown and yellow pigmented scales; the grey region with only black and white scales. Since we cannot resolve single scales, two or more juxtaposed scales of different colours will appear as only the *mixture* of those colours, pooled into a single pixel. Again, we are returning to the methods and successes of the Impressionist artists, who, rather than always mixing paints together to produce a colour absent from their palettes, painted the two colours next to each other as individual dots. Two dots on the canvas become one pixel in the brain.

Taking this idea a little further, what happens when the picture observed is moved in front of our eyes? In this case, we must add time to the equation. We cannot visually process an infinite number of images. The retina is good, but not that good. It is constrained by time restrictions. When an animal moves, a wave of colour flows across our retina. Our visual system acts like a camera on exposure repeat, where photographs are taken in rapid sequence, perhaps at half-second intervals, but the shutter must remain open for a time period during each image-capture. If an object moves during that period, then the image captured will appear as a blur. Or more precisely, each pixel on the photo will reveal the effects of colour mixing where one colour passes over another. And that's what happens to photographs of the blue and yellow macaw in flight. From below, the macaw now appears with *green* patches – the yellow regions of the chest are overlapped with the blue regions of the tail. The longer the eye's 'shutter' is open, the larger the green patch, since the blue tail moves further into the environmental space previously occupied by the yellow chest. And the eye works in the same way.

Our retina can only capture a certain number of images every second – about sixty in bright light and twenty-four in dim light (in dim light the time must be extended to grab enough light to form an image), a number known as our 'flicker fusion rate'. So if the human eye did involve a shutter, which it does not, the shutter would remain open for a sixtieth of a second in bright sunlight. And then, if during that time a wave of blue passed over cone cells previously recording yellow, the pixel formed in the brain would be green. Again, 'green tree frog' syndrome, except here the reason for colour mixing is different – it is 'time' rather than 'space'.

Adding a time factor to vision makes the colours of animals that move quickly more interesting. A complicated, pigmented pattern on the surface of a butterfly's wings may be rarely seen during flight. We see a blurred image of the pattern because the wing passes through a partial closure during each image captured by our eyes. Maybe, then, butterflies' wing patterns evolved so that their colours are only ever observed as they mix together, in which way a conspicuous blue and yellow pattern can become a camouflaged green. But there will always be times during the flapping cycle when the full pattern is observed – at the upper and lower limits of the wing beat. At these times, the wing is perfectly still, and the butterfly momentarily stalls in mid-air.

A demonstration of this effect can be found when a pencil is waved backwards and forwards quickly in front of our eyes. We recognise the pencil only when it reaches the limits of its path, and stops to change direction. For a moment it is still. This same phenomenon can be seen also in peacocks during their displays.

Pre-twentieth-century artists in India, lucky enough to live natively with this pheasant, did not under-represent them in 'gouache' paintings. Other than Indian Yellow, gouache consisted of ordinary pigments. Peacock tail-feathers are structurally coloured, so their displays were always destined to appear less brilliant on paper. The painted peacocks would never attract a peahen in any case – through an eyespot inadequacy.

Look carefully at a peacock when its tail-feathers are raised and more will be seen. During this mating display, the peacock not only shows off its collective eyespots, it also shakes them. The tail-feathers are shaken rapidly, to the point where the false eyes are observed only at the extremities of their movement, while they are still. And this effectively doubles the number of eyespots, which may be no coincidence. The female peahen is the target of the peacock's display, and she is impressed by eyespots. Maybe the male shakes deliberately to double his luck.

Returning consciously to the premise of this chapter, is colour mixing through movement the answer to the milk snake problem?

The model – part one

I placed my order with my university department's workshop – one wooden disc, about the size of a dinner plate and two centimetres thick, with a bolt through the centre. Not the greatest order ever received. Nonetheless, just right to test colour mixing in this case, having the thickness of a milk snake.

I painted bands of orange, black and white along the edge of the disc in the order and width that they occur on the Honduran milk snake, so that the milk snake pattern could be seen from an edge-on view of the disc. I borrowed Adrian Thomas's portable motor, which he had borrowed some time ago from a molecular biology lab (it was designed to power an automatic stirrer). Research happens through punctuated funding – periods of large grants punctuated by skimp-and-save tactics. The milk snake study happened during a thrifty year – my experiments do not become any more sophisticated, as I will continue to explain.

Figure 7.2 Part of a wooden disc with the colour pattern of the Honduran milk snake painted on to the side, taken from video stills while stationary (above) and spinning fast (below). When fast, the individual stripes and colours are not detected; only a blur of the mixed colours is seen (a dull pink in this case).

As soon as the paint was dry, the bolt in the centre of the disc was held in the jaws of the motor and the speed dial was turned to 'fast'. As the disc began to spin, I videoed the coloured edge with an ordinary, cassette-type camcorder.

As the disc first began to turn, the individual stripes could be discerned – orange then black then white then black then orange. On reaching full speed, the stripes could no longer be seen. Now only a continuous dull pink colour was observed, as if the whole disc had been painted with that hue. The video had remained running the whole time, and the video footage showed exactly the same effect – the individual bands were evident at a slow speed, and only a continuous dull pink at a fast speed. This is the effect of colour mixing. Orange plus black plus white equals dull pink, and here was the proof.

This experiment served to rule out colour mixing through movement or indeed via any other means as a cause of the green effect in the Honduran milk snake, if indeed there was a green effect at all. Oh well, so much for the simple solution. So what next? Maybe an examination of the foundations of this problem is needed, and in particular why milk snakes came to evolve this colour pattern in the first place, and what purpose it may serve (although this is not the last to be heard of the spinning disc). This brings me back to the subject of mimicry.

A tale of *two* snakes, or *three*

Because of their extremely conspicuous coloration, a group of snakes in the Americas earned the title 'coral snakes'. Milk snakes were once thought to be coral snakes. In the early half of the nineteenth century, it became known that some coral snakes were venomous while others were harmless. Then came a breakthrough in the form of Henry Bates, working in the land of coral snakes. Bates studied butterflies and, as discussed in the Violet chapter, he uncovered the concept of mimicry in Brazil.

Mimicry provided an obvious explanation for why the bright warning colours of some coral snakes were honest signals of their venom, but those of other coral snakes were deceitful. Quickly, the physical characters of the coral snakes – characters other than colour – were examined and the snakes were placed into related groups. As anticipated, colour provided no indication of kinship, and two fundamental groupings emerged – the venomous snakes on the one side, and the harmless snakes on the other. The most obvious character used here belonged to the fangs – snakes with smooth fangs were non-poisonous, while grooves in fangs always connected to poison glands, indicating the venomous types. The groove character, and its associated gland, is the result of a long line of genetic mutations, or extensive evolution. Colour patterns, on the other hand, were the subject of much fewer mutations, and so could evolve more rapidly (in evolutionary time). The harmless snakes, with smooth fangs and so all related, were given a new name – milk snakes.

The colour patterns of coral and milk snakes certainly do go against the relationship grain. Although most of these snakes contain alternate bands of orange (or red), black and white (sometimes yellow), the width of each band, and the order in which they occur in the repeating sequence, varies with species. Clearly, the milk snakes are mimicking the coral snakes, to gain protection from their warning colours. A certain species of milk snake will have evolved the pattern of a certain species of coral snake, one that lives in the same area. The milk snakes are 'Batesian' mimics (in mimicry the original species is known as the model, and the species that evolved to copy this as the mimic). But how did the *coral* snakes' warning colours, the model, evolve in the first place?

Snakes related to coral snakes exist *outside* the Americas, in the Old World, and they have not evolved such bright colour patterns. Like the coral snakes, these Old World species are also mainly snake-eaters, so the orange, black and white 'banded' colour pattern of the New World species probably has nothing to do with diet. It is also interesting that this banded pattern is confined to snakes of around a metre in length, and no longer. Since this colour combination appears very conspicuous in any environment, there is no question of camouflage, and then the snakes themselves are not nocturnal and do not hide to keep out of sight. Is there something strange and peculiar to South America that is responsible for the banded pattern? On reaching this question during the history of coral snake science, it was discovered that the banded pattern deterred at least some of the snake's enemies, including small, carnivorous mammals (which are heavily dependent upon snakes for food), birds of prey and other snakes. It is likely that coral snakes, and their patterns, evolved from the black and white banded vipers of the Americas. Maybe the addition of orange provided a successful warning signal. But one factor has been overlooked in this discourse so far, and that is the poisonous effect of the coral snakes on their predators.

In the Violet chapter, butterflies were the animals considered as models and mimics, and birds as the receivers of signals, in a system of Batesian mimicry. The butterflies were more numerous than the snakes considered here, and the models were distasteful rather than harmful. Bird predators could *learn* to avoid butterflies with a colour pattern associated with a bad taste through trial and error, and the sacrificing of individual butterflies in the process. In the words of Oscar Wilde, 'Experience is the name everyone gives to their mistakes' (*Lady Windermere's Fan*, 1892). So before long, living alongside the butterflies, existed bird predators with 'experience' – experience to avoid the colour pattern that signalled distastefulness. This pattern had become a 'warning' colour.

The warning colour signal became a selection pressure for a tasteful butterfly species to evolve similar colours, and so side-step all those birds with experience. Too many mimics compared to models and the system breaks down – birds would learn that *most* butterflies with this particular colour are tasty, and so an attack becomes worthwhile. Indeed, when mimics are present in the local community to any degree, the danger to all species with their colour pattern increases. In fact the numbers of models and mimics that coexist are finely balanced, forming interesting mathematical problems for biologists. It is easy to cry 'mimicry' where there is an apparent model and mimic, but extremely difficult to prove it. A combination of tricky behavioural experiments and mathematical probability analyses are required, preferably in addition to a genetic understanding of the mimetic appearance ('phenotype') and the visual response and behaviour of the receiver animal (the predator).

How mimicry, like warning colours, can evolve in the first place is particularly interesting. Remember, for evolution to take place, a genetic mutation happens in one individual, and that individual survives to mate and pass on its new genes to its progenies. Generation by generation, these genes become more and more prevalent, maybe changing further *en route*. But questions remain, because this theory holds only if the mutant individuals survive, a fact that should not be taken for granted. Without the protection offered by camouflage, 'How can a small population of mutants that appear *conspicuous*, which have evolved from a camouflaged population, survive predation to become an independent, viable group in their own right?' Or putting it another way, 'Will the numbers of mutant individuals ever rise above the sacrificial threshold required to impart predator experience of a warning colour?' And then equally challenging for mimicry is 'How closely should a mimic copy the model?'

When prey are confronted with real predator diversity, the cost of complete mimetic accuracy may be too energetically expensive or detrimental to reproduction for the benefit it brings. And in any case, total accuracy may be excessive. Remember the lessons learnt so far in this book. Sometimes corners can be cut in one's visual appearance while still achieving the effect of a more sophisticated façade in the eye of a predator. Alternatively the *general* appearance of a mimetic species may serve as crude mimicry of *several* model species. Living proof that imprecise mimicry works can be found in hoverflies.

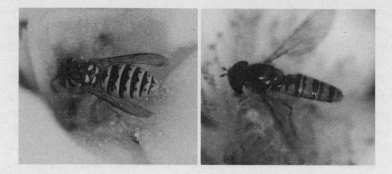

Figure 7.3   A wasp (left) and its hoverfly mimic (right) – far from identical, yet this mimicry is a successful strategy for hoverflies.

Hoverflies mimic wasps, but are rather inaccurate mimics. They are, nonetheless, successful animals. They have achieved the most appropriate body patterns to mimic wasps, *all things considered*. One of those things may be sexual selection, particularly mating colours. When hoverflies outnumber wasps, birds will be less averse to attacking insects of that general appearance. Then mating in hoverflies becomes an issue – the survival strategy turns to species safety in numbers, and to improve the chances of finding a mate, hoverflies must stand apart from wasps and hoverflies of different species. No chance of a wasp mistaking a hoverfly for one of its own, however. Recently paper wasps (*Polistes fuscatus*) have been shown to recognise *individuals* of their own species based on their unique colour markings, like identity tags. Still, mimicry continues to work for the hoverfly against bird predators to some extent. But we must always bear in mind that, in the hoverfly's favour, the potential danger to a bird attacking a wasp is so great that the resemblance does not have to be close for the defence to work. Now I am beginning to demonstrate why mimicry is such a complicated subject. Each case is unique.

So in Batesian mimicry, the target predator learns that a colour pattern is to be avoided through its mistakes. The problem in the case of the coral snake is that the target predator has no chance to learn – of the coral snakes considered so far, once the predator has been bitten, that's the end. As a mark of the potency of their venom, a mouse will drop dead on the spot if struck. So a predator learns nothing from an unsuccessful attack – it will not encounter another situation where it might recognise and avoid a dangerous adversary. On top of this, there exists no evidence that the snake's predators can recognise the harmful nature of a coral snake without prior experience. So all in all, the coral snake's bands cannot constitute a warning colour. Since the milk snakes have evolved an identical colour pattern, what, then, is its purpose?

Well, there are different varieties of coral snakes, and those considered to here were actually the last to evolve. The first coral snake did not evolve with deadly venom, but rather with a mild poison. An animal bitten by this snake *could* recover. Such a victim would then stay well away from snakes of this species, identified by their conspicuous hues – it will have gained 'experience'. Then *two* events in snake evolution took place that involved this warning colour. One, we know, was the evolution of milk snakes, Batesian mimics of this type of coral snake. The other was the evolution of the first type of coral snake considered here – the deadly variety. This must represent a different form of mimicry. There are, in fact, other forms recognised, the most famous being Müllerian mimicry.

In addition to his collection of models and mimics, Henry Bates found that occasionally two *inedible* and unrelated butterfly species were amazingly similar in appearance. This baffled him, but all was made clear a few years later, in 1879, when the German zoologist Fritz Müller analysed *his* results from Brazil.

Like Bates, Müller also assumed that animals learn to recognise inedible prey after bad experiences. Again, this means that the inedible species must sacrifice a number of its individuals to provide predators with experiences. Now, if there were *several* inedible prey species, a predator would have to learn to recognise each separately. But if the different prey species had the *same* appearance, and the predator was unable to distinguish between them, then only that one appearance would have to be learned. In this way the individuals sacrificed to predators would be spread over all species with this appearance, or warning colour. And the more species involved, the fewer the losses to each species. This is Müllerian mimicry.

A good example of Müllerian mimicry can be found in unrelated

South American butterflies sharing the same geographic range. The upper sides of the wings of *Mechanitis lysimnia*, *Lycorea halia*, *Heliconius eurate* and *Melinaea ethra* all have the same colour patterns involving yellow, red and black stripes. This is an honest signal of the unpalatable chemicals contained in all the butterflies' bodies.

Müllerian mimics, conversely, may become models for Batesian mimics. In the South American example above, two further unrelated butterflies – the females of *Perrhybris pyrrha* and *Dismorpha astynome* – occupying the same territory have indeed involved the same yellow, red and black colour pattern, except that this time both species are without chemical protection. Two Batesian mimics have exploited the warning colours of the Müllerian mimic butterflies independently. Well, if one species has been successful at achieving protection through mimicry, then why not two? Again, provided the total mimics do not greatly outnumber the models, the system can be harmonious. But some butterfly species have side-stepped this population size constraint by evolving different colour morphs – some individuals with mimetic colours, others with a different colour pattern and survival strategy. Conceivably, competition should exist between the two Batesian mimetic butterfly species, although this has yet to be tested. Fortunately we can leave such complicated matters in our wake because the coral snakes, with their matching colours but two forms of toxicity, are not examples of Müllerian mimicry. They belong to a third class of mimicry – Mertensian.

The protagonists of mimicry did not have things their own way. Quite a debate raged for many years over the legitimacy of their claims, particularly in the early twentieth century. It was a nice theory, but just that, thought some biologists, like William McAtee. McAtee insisted that proof was needed before mimicry could become more than a convenient explanation of a hole in the theory of evolution, based at that time on a few observations and results. Specifically, he required proof that predators really do what it had been implied that they do – learn to avoid certain colour signals. He even went as far as to demonstrate the opposite.

McAtee made over a quarter of a million prey identifications from 80,000 stomachs, of more than 300 bird species. He found a diverse range of insects represented in his survey area, and claimed that birds ate whatever insects became available to them, regardless of their colour patterns. So insect colour patterns could have no warning role.

It was not until the mid-twentieth century that biologists realised McAtee's mistake in full, namely that he had considered only the end *product* of bird feeding. Behavioural scientists such as Niko Tinbergen studied the product and the *process* of mimicry, and only then could results be placed in perspective. Tinbergen found that predatory birds forage for food in different ways, and that mimicry targets only some feeding strategies. So if the whole range of strategies is considered, as they were in McAtee's studies, then we should expect to find all kinds of prey represented proportionately and learn nothing of how effective the defences of each prey insect may be. Nevertheless, in his time McAtee caused an uproar in scientific circles and the big names in biology clubbed together to defend the ideas of Bates and Müller. One by one the doubters of mimicry began to convert. By chance, these included Professor Mertens.

Mertens, a prominent reptile biologist, was first an opponent of mimicry in coral snakes, until he examined the facts in the field. Then he became a supporter, and proposed a variation on the known mechanisms, which *can* explain the coral snake scenario. Mertens' scenario holds that the most offensive species kills its predators, ruling out the chance of learning warning colours in that case. But in the reverse of Batesian mimicry, this most offensive species is in fact a *mimic*. The model is actually a *less* offensive species with the same colour pattern – a species with undesirable properties yet which does not kill its attackers. This is the only explanation, not dependent on innate recognition, for warning colours in animals that kill attackers – a non-lethal model species with the same colours must coexist. As for Batesian and Müllerian mimicry, in Mertensian mimicry coloration should be regarded only as a particularly effective training device.

So we have two extreme forms of snakes, non-venomous and deadly, that are completely dependent on the intermediate forms, the mildly poisonous coral snakes, as models. Now we can return to the story most relevant to this chapter, where *mildly* poisonous coral snakes are concentrated in Central and South America.

After narrowing down the correct behavioural mechanism in this group of reptiles, we can safely assume that the cohabitation on a global scale of milk snakes with coral snakes is further evidence that the milks are cashing in on warning colours – they are *Batesian* mimics of mildly poisonous coral snakes. This also explains the size restrictions for milk snakes – coral snakes are all around a metre in length. In fact there is an exception to this size rule, but one that also conforms to the theory. One species of milk snake grows much beyond one metre in length. As a juvenile, nevertheless, while still less than one metre long, it displays the orange, black and white banded pattern. As it grows further, it turns completely black.

So the coral snakes were a necessary prerequisite to the evolution of milk snakes. This is a case of colour, or rather the vision of predators, driving evolution. Recently, more's the pity, the distributions of both a coral snake model and a milk snake mimic were studied in detail, and a problem to this tidy theory emerged.

One species of milk snake does indeed live side-by-side with its coral snake model. Here it enjoys the protection deriving from the coral snake's venom, since both snakes share the same predators. But, and a rather large but at that, the milk snake is found also where coral snakes are not. The same predators also exist in this milk-snake-only territory, which stretches way beyond the coral snake range. The question is, 'How can milk snakes, so precisely adapted to coexist with coral snakes, live without their models?' Milk snakes, it would seem, would never have evolved in the first place if coral snakes did not exist, so what turn of events has led to the spurning of mimicry in one of the world's classical mimics?

Since vision has been driving the evolution of this group of animals, maybe we should consider the visual interactions of the predator–prey system in more detail, and in particular the hint given by Adrian Thomas at the beginning of this chapter, when he thought he saw green when he observed a milk snake moving. But first I should introduce another animal that mimics a banded snake, because in this case its colour is well understood. This animal is also coloured in bands, although this time the bands are alternate black and white only. It just so happens that this animal additionally holds the title of 'master of mimicry'.

The mimic octopus

Roger Steene and Rudie Kuiter are renowned underwater photographers. Marine natural history magazines and books are packed with their pictures. And sometimes their continual trawling of the sea floor in shallow, tropical regions uncovers something new to science. That's precisely what happened during a photographic expedition to Indonesia.

Roger and Rudie arrived in Indonesia to capture on film the shallow water marine life in general. There would be no bias towards habitat type, such as coral reefs, and no target animals. During a dive on a sand and mud sea floor, they cracked the camouflage code of a flatfish, perhaps noticing the eyes protruding from the sea floor, and proceeded to take a photograph. Flatfish began their evolution when a group of fish became progressively flattened from side to side, and sought concealment by lying on one surface on the sea floor. During this succession their head twisted 90° so that with one side of its body downwards, *both* eyes projected into the water, rather than one staring into the sand.

The flatfish was approached carefully, as these fish are renowned for their fleeting escapes when disturbed, undulating their bodies to skim the sea floor while kicking up a sediment cloud that becomes a smoke screen. A photograph was taken, but the flash startled the animal, which duly entered escape mode. All, however, was not as it seemed.

As it power-glided over the sand, the flatfish transformed into an octopus. Roger and Rudie had been observing an octopus all along. A magnificent case of mimicry, they thought, so accurate that they were genuinely fooled. This was a small octopus, sixty centimetres from arm tip to arm tip, and prey to a number of visually guided hunters. Such mimicry could serve two purposes – first camouflage to avoid detection in the first place, then, if the cover is blown, to become deceitfully venomous. The flatfish model was the banded sole, a native fish with poison glands at the base of its fins that cause lockjaw in its attackers. But this mimicry was extraordinary because not only had an octopus taken on the shape of a flatfish, a notable feat when the contours of these two animals are contrasted, but it had copied the colour and the behaviour of the flatfish too.

This finding alone was worthy of a report to the scientific world, as for a start the individual had changed from its customary appearance to become a mimic, and such selective mimicry is rare. An animal is generally either a mimic invariably for a stage of its life, or it is not a mimic at all. Then something *truly* remarkable happened. Roger and Rudie traced the path of the octopus in the hope of photographing its other phase, the 'octopus' form, when they encountered only a sea snake.

The banded sea krait is a common sea snake with black and white bands throughout its body length. Like all sea snakes it looks like an ordinary land snake except that its tail end is flattened from side-to-side to aid swimming. This snake is loaded with strong venom, although the effect of this on its predators is not known (another problem for the biologist interested in warning colours – the simple, theoretical solution is that only a small dose of venom is injected into an attacker so that it gains 'experience' but does not die). A closer inspection of the snake revealed a bulge in the centre of its body, as if it had swallowed a large meal. But the bulge possessed eyes. Roger and Rudie had recovered the octopus after all.

This time the octopus had placed six of its arms down a hole in the sandy sea floor, and was waving its remaining two to look like a banded sea krait. Again, another convincing act of mimicry, also involving a major adjustment of the octopus's behaviour, as well as its shape and colour. The flatfish presented thin, pale brown stripes running across its body, while the sea krait's stripes were broad black and white bands, yet both colours had been copied successfully.

This is by no means the end of the story. Roger and Rudie continued to track the same octopus for some time and noticed that it took on a whole host of strange postures, which they photographed. Returning from the field, they couldn't wait to develop their photos and show the scientific world. And there's only one person you go to when you have an octopus poser in the Indo-Pacific – Mark Norman.

Mark Norman is an expert on cephalopods, the group of molluscs containing octopus, cuttlefish, squid and *Nautilus*. He is based at the Museum Victoria in Melbourne. Working on such a visually mediated group of animals, with excellent eyesight and colours to complement, he is obliged to carry a video camera during fieldwork. I could have sat for hours watching examples of his video footage, particularly the octopus that travels over land between lagoon pools, and appears more alien than *Alien*. Mark has also written 'the book' on the cephalopods of Australia, with sixteen different species in Port Philip Bay alone (that's the rather vast but shallow bay of Melbourne). Not surprisingly Roger and Rudie approached Mark, who was interested to say the least, but at the same time a little sceptical. Cephalopods were renowned for their intricate and acrobatic courtship displays, so some of the postures of this small Indonesian octopus could have been misinterpreted as mimicry. Still, Mark wasted no time in organising a field trip to Indonesia, and he took the big guns of underwater filming with him – a team from the BBC Natural History Unit.

Roger and Rudie had pointed to the spot on the map where they made their discovery – the Lembeh Strait in northern Sulawesi. Mark's colleague Julian Finn set off to Sulawesi ahead of the BBC film crew, spending three weeks with local divers. But after more than fifty dives he conceded defeat. Not one individual of the octopus was encountered. And things got worse. Mark joined Julian for a further two weeks to notch up over 200 dives in total, again without a trace of the elusive creature. They were feeling disappointed and embarrassed – nothing to justify their museum's spending, and nothing to show to the BBC team, who by this time were on their way from England.

Not ready to give up just yet, Mark and Julian chartered a boat, large enough also to host the film crew and their equipment, to travel down the east coast of Sulawesi, where Rudie thought he had seen this octopus on a previous expedition (but without photographic evidence). This was a long shot, but they had to do something, although after a week of diving three or four times a day they were beginning to wish they had tried something else. On the evening of their third last night the situation looked worse than ever. The TV producer had become depressed and Mark was resigned to Plan B – how to make a wildlife documentary without the star. And almost as if the octopuses had heard the word 'star', as Mark jumped from the boat the following morning, he virtually landed on top of one.

On resurfacing the film crew could not fail to misinterpret the frenetic, arm-waving Mark Norman as he screamed through his snorkel. Lens caps were popped like champagne corks and the cameras plummeted to new depths (or more accurately shallows – this water was just a few metres deep).

Meanwhile the octopus remained rested in the mouth of a large worm's burrow. All its arms had reached down the hole and only its head and stalked eyes, each topped with a small horn of skin, jutted out from the sand. It watched at first cautiously, and then with probable amusement, as arms, legs and cameras broke the water's surface and sank haphazardly (the usual grace of a dive team's arrival), before floating and recovering their composure. Soon the cameras were rolling and, satisfied the human buffoonery posed no threat, the octopus prepared for its show, and a full performance at that.

The thin, banded arms of the octopus slid out of the burrow and spread over the sea floor. Soon the whole body was revealed, and began foraging over the sand. The octopus's arms appeared rather curled, working the holes in the sea floor with their fine tips, while jets of water ejected through its body funnel to propel it along. The octopus now appeared generally brown – dark brown bands separated by pale brown regions. Actually this was the animal without one of its supposed masks. Then the show-curtain rose. 'Action.'

The octopus drew in its arms around its body to reveal a flat, teardrop shape. Its pale brown regions bleached to reveal the now lucid dark brown stripes traversing its superficial body (its body *and* trailing arms). Eyes were raised fervently so that they protruded high into the water; the funnel poked out ready to jet-propel the animal over the sea floor. And off it went, rippling its body in the manner of a banded sole, which must undulate to move. This was the flatfish mimic that Mark had hoped to see, and he and all the film crew acknowledged the close resemblance to the fish. This was no coincidence – it was almost certainly mimicry.

The octopus returned to its original burrow.

Next on the programme was 'large solitary anemone' (*Megalactis* species). Arms were all raised over the head, transformed entirely to very dark brown, and each bent into a zig-zag shape, curling inwards to some extent. The large anemones in this region look just the same, except that unlike the octopus they are armed with powerful stinging cells, strong enough to kill fish.

Soon another octopus had been spotted, then another, then another (totalling one male and three females altogether). They were all observed closely and filmed enthusiastically. Act three was the sea snake disguise, and appeared just as Roger and Rudie had reported. This, thought Mark, was the most convincing mimicry performed by this animal so far, until one individual opened the lionfish act.

This octopus swam above the sea floor with all arms spread around its body like the poisonous spines and banners of a lionfish. Lionfish are those graceful creatures often seen floating mid-water in tropical aquaria, with striped bodies at the centre of a mass of banners radiating out from the body into all directions. They can also be found in those 'deadliest animals' picture books. Their banners, each as long as the body itself, are the remnants of side fins. They are all banded in black and white throughout, and each is made taut by a poison-tipped spine running along its length. The flared arms of the octopus, meanwhile, with clear black and white bands, made for a rather convincing lionfish double. But this mimicry was especially persuasive because it involved the octopus being prised from sea floor and into open water.

Mark has witnessed many species and individuals of octopuses in the wild, and the one thing they generally do *not* do is venture into open water. They tend to reside on the sea floor. On top of that, they spend all their time hidden under rock crevices or camouflaged, and so invisible. Flaunting oneself on the sea's centre stage – in mid-water – and waving conspicuous arms for all to see was a most risky strategy for a defenceless creature. And all in broad daylight, in the presence of many potential predators with sharp eyesight (and teeth). The octopus's 'lionfish display', and the similarity to a native lionfish, appeared increasingly like no coincidence. The evidence began to strongly favour mimicry. Although rare cases of other octopuses performing mimicry are known, this species became dubbed *the* 'mimic octopus'.

There may be further acts in the mimic octopus's repertoire. Could the initial posture observed, with only head and eyes projecting high from a burrow, be the mimic of a mantis shrimp – a large crustacean

Figure 7.4   The mimic octopus; in various guises (left), with the model species pictured next to each. Top, banded sole (a flatfish); middle, lionfish; bottom, banded sea snake. Photographs by Mark Norman and Roger Steene.

with lightning-fast claws – as it waits part hidden within its lair? Then there is the possibility of a *swimming* sea snake as arms are assembled into two groups of four – one group ahead of the body, one behind – again in open water. In fact this octopus wriggles into other postures that have yet to be assigned to animals or objects.

Of course, most cases of supposed mimicry here need some explanation. The sea snake, flatfish, lionfish and anemone all contain venom that can *kill* their enemies. Are these *model* animals, then, really displaying warning colours and shapes (visual cues), considering the coral snake lessons? Or are their appearances merely incidental? If they are incidental, the octopus's mimicry may have a purpose other than to warn of (fictitious) venom. Instead, the octopus may benefit simply from appearing as something other than an octopus. Maybe this region of water is too perilous for a small octopus, where a host of local predators have efficient means to detect and capture them. Octopuses and their relatives are known to make excellent meals for fishes since they are nearly all muscle ('meat'), have no spines, no armour and very few are poisonous. Maybe an elfin octopus could not expect a long life in this stretch of Indonesian water. A flatfish, on the other hand, may suffer a much-reduced risk of predation. Few predators would attempt to tackle, or be bothered to attack, an animal with a flatfish's size and shape. In which case, an octopus stands a better chance of survival if it can emulate the appearance and lifestyle of a flatfish (comparable to a puffer-fish, which gulps down water to inflate itself and appear too large for its predators' mouths). And the same goes for the skinny sea snake, blubbery giant anemone, and the ball of long spines that is the lionfish – they all suffer generally less from predation than an octopus.

Still there is the question, 'How does the mimic octopus know which animal to mimic in each situation?' But this may be not so challenging, since it will clearly fare better as a different animal to different predators. So the visual identification of a predator may elicit a certain response, one way or another. This whole case of *multiple* mimicry must be driven by a *diversity* of predators.

The really big question so obviously remaining is, 'How did this octopus evolve *multiple* mimicry?' Elsewhere in this book, mimics were born with and died with a single, inflexible act of mimicry (or were invariable mimics for a complete phase of their life).

The first explanation is the conscious one, involving the learning and active selection of behaviour. The octopus may learn its mimicry, possibly while a juvenile, and decide where and when to employ it, and which particular act in its repertoire to employ at that. Its switch to mimicry mode must begin when it perceives danger, through its eyes. It would then assess the threat and consciously choose the most appropriate response. It would gauge whether the predator approaching is more likely to avoid a sea snake or a flatfish. Unfortunately the only attacks on the mimic octopus witnessed by Mark Norman were made by just one enemy – the highly territorial damselfish. In each of the four times he witnessed such an attack, the octopus would always abandon its foraging to mimic a sea snake.

The alternative explanation for how the mimic octopus came to possess such a versatile behaviour is the inherited or subconscious one. Here this multi-mode behaviour is considered the product of a long and gradual evolution, so that the ability to mimic is innate. In this case the eyes would again image the predator approaching, but this time an automatic response would be elicited. The octopus would involuntarily switch to a specific act of mimicry.

Maybe the mimic octopus's early ancestors evolved genes that triggered the comparatively simple switch from its foraging form to a flatfish shape whenever danger was observed (the flatfish shape is the most similar to its regular, foraging form). The progress of this ancestor would then have evolved genes that triggered the subconscious retreat to a burrow, while trailing out two arms, as an automatic response to the sight of a damselfish. But a spanner in the works of this idea is that the octopus's environment may be constantly evolving too. Predators evolve, and come and go to various degrees. In other words, during the mimic octopus's evolution the type of predators that hunt it today may not always have coexisted. That is unless this stretch of Indonesian water has been unusually stable over hundreds of thousands of years.

Another sticking point for the subconscious explanation is that in all the literature we have trawled through on mimicry so far in this book, innate recognition is a subject that has never emerged. On the other hand the mimics covered so far have all gained their guises directly from their genes. But the mimic octopus may be a special case, more closely allied to the target enemies (*receivers*) of other mimics than the mimics themselves – it could use its visual system and make an informed decision. Maybe during its life the mimic octopus is first an effective receiver and *then* a mimic. As Mark Norman suggests, we need to find juvenile mimic octopuses to determine whether they too possess the ability for multiple mimicry from the beginning, or whether they are in the process of learning their future tricks. Still, this is one amazing creature all the same.

Moving on, and back to the path I have prepared for this chapter, the lesson of the mimic octopus most relevant here is how it changes its *colour*. Yes, it mimics a snake, as does the milk snake (with comparative ease), but also the octopus is banded – the milk snake is banded too. And then the octopus changes its colour – another property proposed for the milk snake. As I mentioned earlier, we do know how this is achieved in the mimic octopus. Time to reveal all with NanoCam.

Detail of the mimic octopus's colour factory

As the octopus rests on the sandy sea floor near the entrance of a worm's burrow, at just a metre's depth, its skin is weakly striped with alternate pale brown and darker brown bands. NanoCam is manoeuvred into a pale brown stripe in one of its arms.

Breaking through the surface of the skin, and weaving between the skin cells along its way, NanoCam travels to a region of muscle that works the movement of the arms. White light appears to radiate from this region. Numerous fine muscle fibres are imaged on the monitor as a beam of sunlight blazes through the skin and strikes them. There are collisions all around. Rays of light of every colour randomly bounce from fibre to fibre, like a pinball.

Comparisons to pinball machines have been made before in this book, where light scattering was involved. Again, scattering is to blame, this time for the white component of the mimic octopus's appearance. All those rays that randomly reflect into all directions recombine in each single direction to reform white. A white hue results, spread over all angles, like the effect of a pigment. And satisfied with the simple explanation for white, NanoCam is turned towards something else noticed along its path to the white region – something that, when struck by sunlight, reflected no rays. NanoCam travels back towards the object for a closer look.

The object is a rather large cell, nearly a whole millimetre across. Its thin, elastic outer wall is stretched out. The whole cell takes the form of a disc with several thick, long branches radiating from its periphery to form a star shape. Many thin muscle fibres are attached to the branches and appear to be pulling them, stretching them out – the muscle fibres are contracted. And connected to those muscles are in turn tiny nerve fibres, which themselves plug into a larger nerve fibre that runs through the skin for as far as NanoCam can see.

Inside the cell is a large, sack-like nucleus that contains the cell's DNA. But beyond the nucleus, within the watery cytoplasm, are molecules that are affected by the sunlight. There are lots of them, distributed evenly throughout the cell. NanoCam focuses in further, on to a single molecule.

As a beam of sunlight strikes the molecule, electrons are jolted from the outer shells of some of its atoms. This is a melanin pigment, as encountered in the vole's fur in the Ultraviolet chapter. Its effect, in this case, is to absorb the energy in *all* the sun's rays, all the wavelengths from ultraviolet through to red. The melanin molecules are clumped together to form granules. NanoCam pulls away to once more image the whole cell. Even the cell's branches are noticeably filled with melanin granules. And from beyond the octopus, the cell is even visible to the unaided eye, as a mere dark brown dot. Still, it is visible all the same. The cell is known as a chromatophore, the reason for which will surface soon.

Meanwhile, in the water a damselfish has approached the octopus and incited the sea snake mimicry. NanoCam is positioned within one of the two arms that protrude from the burrow, the burrow that is now filled with the other six arms. Something happens to the chromatophore in view.

The octopus images the damselfish on its retina. Nerve impulses fire and travel through the optic nerve to a large offshoot of the brain known as the optic lobe. From here, the message 'Damselfish at eleven o'clock' is sent to the lateral basal lobe – the central region or 'headquarters' of the brain. The lateral basal lobe makes a decision which it sends to the chromatophore lobe, yet another offshoot of the brain. The chromatophore lobe is the switchboard for colour and mimicry behaviour (actually we can consider colour as an aspect of behaviour). The message it has received in this instance is 'sea floor sea snake', and of all the connections available – foraging, flatfish, lionfish, mantis shrimp, sea snake (swimming or on the sea floor), and maybe others – the 'sea floor sea snake' one is made.

Suddenly, as part of the 'sea floor sea snake' message, the electrical signals feeding the chromatophore in NanoCam's view cease, causing the chromatophore muscles to relax. The contraction or relaxation of the muscles affects the shape and ultimately the colour effect of the chromatophore.

When the muscles attached to the chromatophore are relaxed, the elastic cell wall also becomes relaxed, like a deflated balloon. The chromatophore shrinks and the watery cytoplasm within surges from the tips of its branches towards the centre of the cell, taking the melanin granules it supports with it. The melanin granules are gathered like logs in a torrent and are dumped only when they reach the very centre of the chromatophore, where the cytoplasmic rapids cease. Here they become concentrated into a tiny region, less than a tenth of a millimetre in size. NanoCam takes a step back and reassesses the whole cell.

The chromatophore now looks quite different. Its branches, which were once dark brown, are now transparent. There is a very dark spot in the centre of the cell, but it is so small that now the unaided eye cannot see it. Instead the eye sees only the light reflected from the bulk muscle layer behind it, which is white. The effect of the chromatophore shrinking is that where it once imparted a dark brown spot to the octopus, it now adds nothing to the animal's colour. This cell type is

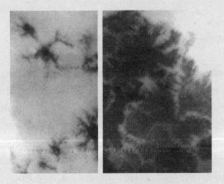

Figure 7.5 Light micrographs of chromatophores of an oar-footed shrimp, *Gloiopotes* sp.; fully contracted (left) and fully expanded (right). These cells are around 0.5 millimetres from tip to tip.

called a 'chromatophore', from the Greek meaning something that carries colour.

The complete band of the octopus's arm in which NanoCam is positioned changed from pale brown to white. The octopus's skin is packed with chromatophores, hundreds of thousands of them, all individually controlled by the chromatophore lobe of the brain. They work in units, where all those within a single band of the octopus's arm are subjected to the same nerve impulses (a conservative approach that keeps the whole colour system practical and achievable within the physical constraints – the total nerve quantity and brain capacity – of an octopus). So a band can change its shade from white, where all its chromatophores are turned off and the white scattering from beneath is observed, to dark brown, where all its chromatophores are expanded to reveal their melanin and block out the white. Or they can appear any shade of brown in between by expanding and contracting their chromatophores to various intermediate degrees. And all in less than a second. That's how the mimic octopus is able to change its colour.

The chromatophore considered here is filled with the pigment melanin, and is called a 'melanophore'. But there are other types of chromatophores in other octopuses, filled with different pigments and so controlling the appearance of other colours.

Chromatophore variety

The fascination in animal colour changes dates back to ancient times, although early explanations focused on an anticipated breakdown and rapid reformation of certain chemical pigments. This null idea hovered over philosophy until the concept of an adjustable cell was realised in the mid-nineteenth century. And as soon as the first chromatophore was unveiled, the search began for their variations and occurrence in the animal world.

Remaining with octopuses for now, the widely distributed *Octopus vulgaris* is normally a brownish-coloured animal. Its skin is filled with chromatophores, but not just the dark brown or black melanophores.

Below the melanophores is a second tier of chromatophores filled with a yellowish 'ommachrome' pigment. Each tier is serviced by its own nerve fibres, so the degree to which the chromatophores of each tier expand or contract is under separate control and can be varied. This means that the octopus can achieve a variety of hues. The expanded chromatophores are those observed by the unaided eye, and their colours mix to produce a single hue, just as the pixels on a television screen work. If the dark brown and yellow chromatophores expand, then the octopus will appear orange-brown in this region.

Beyond the coloured chromatophores in the skin of *Octopus vulgaris* is a third tier of chromatophores with a trick up their sleeve. This trick is to mirror-reflect the colours in the animals' habitat not covered by the chromatophores above, but that's a story for the following chapter. Now we should consider briefly how chromatophores become useful to their octopus hosts.

Recently I observed an *Octopus vulgaris* on the floor of the Red Sea, made obvious because it was totally white (possibly through a reaction to an enemy, which could have been me). Then instantaneously it disappeared. I was left in awe of its swiftness, since I saw nothing of its escape. But before I had chance to turn my head, I noticed it once more – a white octopus, and in exactly the same spot. The animal in fact had not moved at all, but had instantly changed from totally white to perfect camouflage colours and finally back to white again. This was awe-inspiring proof of the effectiveness of chromatophores.

Related to the octopuses are the cuttlefishes and squids, and these are simply spectacular for their chromatophore diversities and colour changes. In an instant, waves of colour can flow across their bodies, and then they may change their entire hue from yellow to red and back again. But at the same time they appear to assess and continually reassess the circumstances in their environment through their eyes. Faced with a cuttlefish in the water, one cannot help but be mesmerised by its large eyes which just ooze intelligence, jutting out from behind the primitively waving tentacles. And the raising of one's arm elicits a response in colour as well as a slight scurry in movement, suggesting a level of sophistication that says, 'I'm in control of this situation.' An aggressive rush towards the animal, on the other hand, is accompanied by a total bleaching of its body and its disappearance from sight. Cuttlefish seem to assess the circumstances correctly.

Squids and cuttlefishes are in fact well known for their language of colour. They use their colour patterns to communicate with each other; to signal a willingness to mate or an unwelcome approach. Biologists such as Roger Hanlon at the Marine Biological Laboratory (Woods Hole, Massachusetts) have mapped the whole dictionary of colour patterns in certain species, and even deciphered their meanings in some, after observing the reactions of the signal receivers. It is difficult even to know where to start in writing about the colours of this group – there are whole books devoted to just that, and rightly so. I know I could never do justice to their colours in such a limited page space, which is why I choose only to acknowledge them.

Animals that possess chromatophores range from insects and worms to frogs and fishes, not to mention the most famous of all – the chameleon. In some cases the chromatophores work in different ways. In crustaceans such as shrimps, the chromatophore itself does not change shape, but rather the pigment migrates freely within the cell. But, conscious that I should return to the milk snake, I must move on and assess the evidence collated so far in the course of this chapter. To begin, do green chromatophores exist in the milk snake, a reptilian relative of the chameleon?

Chromatophores and other possibilities

NanoCam is injected into the skin of the milk snake, first within an orange band, then within a black band, and finally into a white band. Orange and black pigments, and white scattering are observed. But in all cases the pigment and scattering elements are arranged in bulk throughout the skin, not within cells. There are no chromatophores anywhere in the snake's skin. The milk snake does not employ the colour change mechanism of the mimic octopus. So where does that leave us?

Right at the beginning of this chapter, colour mixing through movement seemed the obvious explanation, but that was ruled out during the disc experiment. We know pigments on their own are not able to change the hues of their hosts – they do not have the ability to break down and reform as early biologists thought. Searches for green bioluminescence and fluorescence in the milk snake failed – it did not glow green in the dark or when placed under an ultraviolet lamp. Finally, structural colours were ruled out when Adrian Thomas mentioned that he thought he saw the green also in a *picture* of part of the milk snake's outstretched body – the picture was printed with ink, and so could contain only pigments. And that's just what NanoCam had found too – that the milk snake's skin contained ordinary pigments and no other colour-producing factory. At this stage in my deliberations I noticed something that seemed relevant to the milk snake case in a book of optical illusions.

Among the pages of parallel lines made to look as if they converged, and 'moon illusions' where objects were made to appear either small or large depending on their surroundings, there lay a figure containing two grey squares with small, black crosses at their centres. This figure included a play on complementary colours – those which, when mixed in appropriate proportions using coloured light beams (not paints), produce the perception of grey. The square on the left was uniformly grey, but the square on the right contained a circle divided into four equal segments by thick grey lines that eventually merged into the grey background. The segments were coloured red, green, blue and yellow. The reader was instructed to spend forty-five seconds adapting to the coloured segments in the square on the right by fixating on its central cross steadily. Then one's gaze should be quickly transferred to the square on the left, again centred on its cross.

I followed the instructions and saw just what had been predicted – superimposed on the grey square on the left were illusory coloured patches. And they were *complementary* in hue to those on the right. That is, the blue patch gave a yellow after-image, the yellow patch a blue after-image, the green patch a red after-image, and the *red* patch left a *green* after-image. That's red turning to green. A breakthrough for the milk snake challenge, I thought.

Then I realised that this game only worked when considerable time was spent adjusting the retina to the four-coloured circle, hence the forty-five seconds suggested in the instructions. Adrian Thomas, on the other hand, had first noted the green effect when he observed a milk snake *moving*. A snake moving in front of our eyes would provide nothing like the time required for our retinas to adjust to this trick. Anyway, it was red that had turned to green in the optical illusion book, not orange. Some milk snakes do contain red in their body patterns, and originally I wondered if orange was close enough. But reading around the subject of after-images further, I discovered that strictly speaking the complementary hue for green is a reddish-purple. It seemed that orange was just not right.

At this stage it was beginning to seem that the green colour of the milk snake was a figment of Adrian Thomas's imagination. But before giving up all hope of ever finding green in a milk snake, maybe there was something in the model disc that was worth a second look. It seemed that, although it did not hold the answer to the milk snake 'green' problem, I was on the right track with the after-image experiment and the general subject of optical illusions after all.

The model – part two

Returning to the wooden disc experiment, there was something I deliberately failed to mention when I first described the results within this chapter because I thought *I* had been seeing things. I thought I had noticed a dark green colour as the speed of the disc reached the point where its individual bands could no longer be discerned. But I had checked stills from the video footage of the spinning disc, and it showed nothing of this kind. The sequence of stills from the video recordings, placed in order of increasing speed, showed bands simply fading directly into the continuous pink coloration, with no intermediate phase. I *must* have been seeing things.

After ruling out structural colours, bioluminescence, fluorescence, colour mixing and chromatophores, I was running out of alternatives. While considering my options, I ran the simple disc experiment again.

As the disc increased in speed for its second time, the individual bands were distinguished as they flashed in front of my eyes, until eventually they became impossible to discern. But then I saw it. Green. Definitely, olive *green*.

This time around I took more interest in the speed of the disc. Rather than winding the motor's dial directly to 'fast', I turned it gradually so that I had control over how quickly the coloured bands passed before me. I increased the speed gently until I had reached the point of green, and this time I held it there. Olive green was apparently emanating from the painted part of the disc. As I increased the speed a little, the green coloration remained, until I turned the dial just too far and the dull pink reappeared. I had not only discovered that the green existed after all, but that it was associated with a specific *range* of speeds. Using the circumference of the wooden disc and the time for a notch in its edge to complete ten revolutions, I calculated this range to be from 0.8 to 1.2 metres per second.

Now I was beginning to question, 'Was this green effect due to a problem with my eyesight, one shared perhaps with Adrian Thomas?' 'Or did everyone see green?' I showed the spinning disc casually to four students. They all saw green too. That was something of a psychological breakthrough. For the first time, I could speak about my findings with confidence.

The solution – 'Benham's Top'

My seminar in 2001 at Sussex University was timely. I was due to meet up with resident vision and perception experts Mike Land and Dan Osorio, and the first to greet me was Mike. I dropped into our conversation a description of my recent milk snake experiment and results and – I will avoid beating around the bush here – he had just two words for me: 'Benham's Top.' It seemed that I had arrived at just the right office.

Benham's Top has always been a subject for psychologists and physiologists. Not physicists, since its effect has nothing to do with optics or the properties of light, and certainly not biologists, since it has always been unknown in nature. Simply, Benham's Top is a flat disc divided into a white half and a black half. The black half is featureless, but the

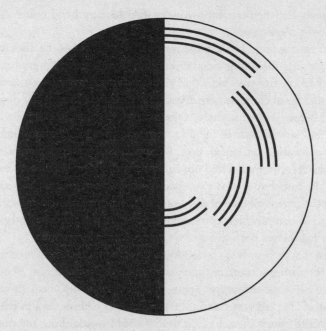

Figure 7.6   One form of Benham's Top. When spun at certain speeds, the outer arcs appear greyish blue, the next group of arcs olive green, the next group mustard yellow and the innermost group a dull red.

white half includes additional black stripes. These can take different forms, one of which is a series of thinly-lined arcs, arranged in groups of four, where each group occupies a forty-five degree sector of the disc, and become progressively closer to the centre. This disc is viewed from directly above rather than edge on.

When Benham's Top is spun at a speed corresponding to just below our flicker fusion rate (the speed of a flickering image at which the eye begins to integrate 'frames' into a continuous image), the black arcs take on colours. The colours are always the same – greyish blue, mustard yellow, olive green and brownish red. They each appear at specific speeds, or rather speed ranges. So as arcs at different distances from the centre of the disc move at different speeds, it is possible to observe all four colours at the same time. Any faster and the greyish blue from the outer arcs is lost as the other colours all move one place towards the circumference. The inner arcs become merged into grey (black and white mixing via movement).

This disc was invented in 1894 by the toy-maker C.E. Benham, although the phenomenon of black and white patterns causing people to see colours was described first by the scientist G.T. Fechner in 1838. Benham called his disc an 'Artificial Spectrum Top', and sold it through Messrs Newton & Co. He did not know why it worked and actually we still do not know today. The colours are pattern-induced, and must have something to do with lateral inhibition – that the eye becomes progressively limited in its vision towards its periphery – and with the different rates at which the different cone cells can fire their signals when all are equally stimulated at the same time (by the *white* 'arcs' on the disc). It is believed that the blue cone cells respond more slowly to light than the green and red cone cells.

The cone cells in the retina convert the patterns of the disc in time and space into nerve firings in time and space. Clearly, as the speed of the disc increases, a point is reached where the two (light stimulus and cone cell firings) do not follow the usual translation. Also, that different parts of the brain are used to interpret different aspects of some patterns may enter the equation. And that's really as far as psychologists have reached and agreed upon regarding the workings of Benham's Top. It is a real puzzle, and may be solved synonymously with the complete understanding of how visual images are processed and interpreted in general.

To demonstrate that Benham's Top is all about our visual processing and not the wavelengths of light that leave the spinning disc, it is interesting to consider a Royal Institution Christmas Lecture from the early 1960s.

The Royal Institution Christmas Lectures are annual science events that are filmed in front of a young audience, to be later shown on British television. In the early 1960s television sets were exclusively black and white, and during this era a Benham's Top demonstration appeared in one of the broadcast lectures.

During filming, the audience in the Faraday Lecture Theatre in

London all witnessed colours from the spinning black and white disc that had been projected on to a white wall. But during the subsequent television broadcast, everyone at home saw the colours too. They saw *colours* on their black and white television screens! This was the first time colour was observed on television sets anywhere in the world, pre-dating the advent of colour television.

Of relevance to the milk snake, I timed a model of Benham's Top and counted how many revolutions it made while green was observed. I calculated the circumference travelled for the precise arcs that appeared green and worked out the effective speed at which Benham's Top causes an olive green effect. Once more there existed a range of allowable speeds, and these corresponded exactly with the speed range I had measured for the wooden disc model, bearing the image of the milk snake's body. A perfect match! Now all that remained was to establish the speed of the milk snake itself, as it moves naturally. Unfortunately this had never been measured.

I scoured the zoos and pet shops in every country I visited for a year, and found not a single milk snake. Several times I ran into 'Just sold out' or 'Waiting for a delivery', but I began to wonder if these really were animals that could withstand captivity at all. I had begun to compare rival air fares for a trip to the snake's native Central America when a contact of a contact heard a rumour of a milk snake held in . . . Oxford.

I was there, with my stopwatch and tape measure, within the hour. The snake was residing in a vivarium in Warwick Vardy's house. Warwick had acquired considerable knowledge of reptiles through his position of zoological technician, and he was more than pleased to help me out. He offered use of the milk snake at his house – a Honduran milk snake at that (with exactly the pattern I had been working on). Although I would have to be quick, since the specimen was travelling to Paris the following day to appear with actress Michelle Pfeiffer in a television commercial for a mobile phone company (I hope the plans were not for an atmospheric *black and white* commercial).

After a year of searching for milk snakes, I seized this opportunity and marked out the snake's sprint track. 'Sprint' soon became the operative word, as this soon-to-be-famous serpent was lightning fast.

Fortunately the narrow track was enclosed by walls, so the snake could neither escape nor change its course, which would have made the measurement of distance, and consequently speed, tricky. Warwick carried the harmless snake to the starting line, and off it went, in Olympic fashion. I was positioned at the finishing line to turn the milk snake around so that it could start all over again. And I didn't lose the animal once (Warwick may not confirm this). Overall everything went to plan. Not only did I time several sprints, but I also noticed the green coloration as I stood over the snake when it left the starting blocks.

The times recorded for the twenty-metre dash varied to a degree, but the average speed lay within the range for 'green' as measured from Benham's Top and the model disc. Remarkable! That was the final piece in the puzzle. Now I was satisfied I had an answer to the milk snake's 'green'.

The milk snake did go on to become a star of the screen in Paris. But sadly, following its first brush with fame, it was never seen again. It did not return to Oxford and Warwick Vardy. I hunted long and hard for a second milk snake specimen to conduct further time trials, but my efforts went unrewarded. I had to be content with what I had, which was nice evidence all the same.

The solution to the problem posed at the beginning of this chapter now appears to lie in 'Benham's Top'. It would appear that the moving milk snake appears olive green by the same phenomenon, albeit we do not yet understand precisely what that is. The milk snake also contains moving black and white bands, and appears one of the four colours in Benham's Top – olive green. Not pale green, turquoise green, lime green or racing green, but very definitely *olive* green.

The milk snake may even add something to our understanding of Benham's Top. The snake not only has black and white regions but also an orange component. Yet the phenomenon still works. Could this tell us something about the way we see the moving hands – about how our cone cells are firing? Why does the addition of orange apparently make no difference, at least to the olive green colour? Then again, maybe the orange serves no visual purpose when the snake is moving fast. Well, this is not my subject and not my problem. I am just pleased to have come through *my* particular challenge in one piece, with the help of my colleagues. And the first finding of Benham's Top in animals is the icing on the cake.

Another weapon in the milk snake's arsenal . . . and maybe that of other animals?

Conceivably, the production of a green colour over the whole of the milk snake's body, as it moves naturally, plays a part in the animal's behavioural strategy. While moving through greenish grasses, it would appear camouflaged. This would represent one of the most astonishing cases of *camouflage* known – the ability to switch from mimicry to camouflage with little adaptation of one's behaviour or nerve activity. It is almost lazy camouflage, cheating the long line of evolution usually required in the descent of animals. And what a great behavioural strategy. To be camouflaged when moving normally, but if spotted to switch to Plan B – simply stay still or move more slowly and coral snake mimicry is achieved. Wonderful!

Unfortunately we cannot yet confirm this idea beyond all reasonable doubt because we do not know the precise flicker fusion rate of the predators of the milk snakes, or more precisely whether the predators see the green effect too. Their predators are the same as those for coral snakes – small, carnivorous mammals, birds of prey and other snakes. The most significant among these, however, is not known.

Most of the work on flicker fusion rates ended in the early 1960s before the subject of its variation in different animals could be properly explored. Maybe the snake predators and mammal predators of the milk snake have a flicker fusion rate similar to our own (although the mammal predators see using only two visual pigments or cone-types). In this case, they would see green on Benham's Top as we do, and a milk snake at full speed would appear green to them too. Then there is also a *range* of speeds at which this effect works on us, corresponding to an allowable variation of flicker fusion rates while still perceiving green (although the effect of background light conditions, in particular those in the milk snake's natural environment, has yet to be tested). It does appear likely that the milk snake's Benham's Top effect is not incidental but the product of evolution. Although conceivably very little evolution, since only a minor adjustment in behaviour is required, where the milk snake must travel within a certain range of speeds.

The bird predators of milk snakes are known to have a much higher flicker fusion rate – 170 Hertz is the highest recorded for a bird. But that's not to say that Benham's Top definitely doesn't work for birds. Still, even if it worked only for some of the snake's main enemies, then that may be enough to explain the employment and evolution of this effect in the milk snakes. No animal species, of course, has absolute protection. But a relative degree of protection, which even slightly reduces the number of predators, is of great advantage.

The Benham's Top effect and the fact that the milk snake appears olive green when it moves may explain why the milk snake is sometimes found where coral snakes are not, providing the green does play a camouflage role. Maybe the employment of camouflage alone is enough to ensure the survival of these animals, although better protection would be sought from the additional mimicry. But that's beginning to stretch the speculation a little too far.

In the future, the colours and speeds of other milk snakes and coral snakes will no doubt be examined for the Benham's Top effect, since there is some variation in the width, hue and sequence of the coloured bands. And then does the Benham's Top effect work on animals with only black and white stripes? That would be an interesting thought, opening all sorts of doors. There exists a tantalising diversity of animals with black and white stripes that move at appropriate speeds, from butterflies to zebra. Come to think of it, zebra do appear as a continuous *mustard-yellow* hue when they gallop . . .

Whatever the outcome of future studies in this area, Benham's Top is now officially a cause of colour that can be found in animals, function or no function.

A tonic for Darwin

Again cases of mimicry have been encountered in this chapter where a predator's eye is tricked. In the case of the hoverfly even very inaccurate mimicry works, highlighting further that the eye, or visual system as a whole, is far from perfect. It is not worthy of the title 'perfection' as Darwin so anxiously proclaimed. If Darwin had known of the hoverfly and wasp association, and further still the rather crude mimicry of the mimic octopus, he may have thought twice about singling out the eye as a spanner in his master works.

Visual illusions also fool our visual systems, often in spectacular fashion, but they are generally not natural. So it is unfair to criticise the eye for its shortcomings in these cases. But Benham's Top may prove an exception. If the milk snake is really using the Benham's Top effect to appear camouflaged, then this represents yet another tonic for Darwin's anxiety. Milk snakes are conspicuous orange, black and white banded snakes. If their main predators look at them as they move through the grass and see only a field of continuous green, then the snakes have achieved camouflage through weakness in their predator's vision. In which case the predator's eye *is not perfect after all*.

The variety of colour factories in nature has now been all but exhausted. Movement, coupled with the psychological interpretation of colour that does not really exist, represents the last serious option on nature's palette. The next chapter, covering the final colour in the visual spectrum, will reveal appealing variations on one or two of those other options covered so far. As a matter of interest, it will also disclose a most curious factory for a colour display that may or may not deserve a place on nature's palette. But on a subject I could not possibly over-look, the next chapter will travel to a world not yet embraced in this book. That is the least known of all Earth's environments – the deep sea.

red

The problem:
Why are red, oar-footed shrimps seen by some predators in the
deep sea, where light is known to be only blue?

In the sea, animals face a different challenge in their adaptation to
vision. At the very surface, the situation is much the same as on land,
but on submerging deeper, the sun's spectrum begins to change – it
becomes shorter.

Moving downwards from the surface, red is the first colour to drop
out of the environment by perhaps ten metres. Red is followed by
ultraviolet, orange, violet then yellow. Water molecules absorb the rays
corresponding to these colours. It is interesting to carry a compact disk
on a deep dive. The disk bears a diffraction grating that is fuelled by
sunlight, hence the reflected spectra. It splits up the sunlight into whatever colours are contained within, and by around ten metres depth the
compact disk no longer reflects red. At twenty metres, violet and
orange are notably absent from the spectra too.

At around 100 metres depth, only the blue and green parts of sunlight remain. Then, falling to around 200 metres, the green drops out,
leaving the environment with only blue. At 200 metres depth a compact
disk's spectrum would contain only blue. This situation exists to about
1,000 metres depth – from 200 to 1,000 metres in the ocean, sunlight
is exclusively blue (the precise depths do vary on the conditions of
both the sunlight and the sea water). Julian Partridge of Bristol

University embarked on a special trip in a submersible to photograph this part of the ocean without using a camera flash. After much explanation of his methods, he exhibited his result at a scientific conference – his thirty-five-millimetre slide lit the projector screen blue; pure, uninterrupted blue. Although this experiment might seem somewhat unchallenging, and was received with the humour with which it was presented, it was actually good to know. The blue photograph confirmed the theory.

The deep sea floor continues to a depth of around 6,000 metres (excluding further trenches), but sunlight ends altogether at around 1,000 metres, at least to the level where it can be used for vision by even the most sensitive eyes. I will term the zone of blue sunlight, between 200 and 1,000 metres, the 'mid-deep sea', and beyond 1,000 metres the 'deep sea'.

In the deep sea there may be no background illumination provided by sunlight, but there is vision. Animals exist with eyes even here. There *is* light in the deepest water after all – bioluminescent light.

Although without a continuous light field as on land, darkness in the deep sea is at times broken by patterns of incessant flashes and streaks of light. Around 80 per cent of deep sea animals can make their own light – vision remains a powerful sense to animals even where there is no sunlight. And it has long been known that the bioluminescence of deep sea animals is blue.

It is equally recognised that eyes in the deep sea detect only blue light too, as one would expect. There is no purpose for yellow or violet cone cells in retinas here because there is no yellow or violet to see. Even if deep sea animals contained yellow and violet, or orange and green pigments, these colour factories could play no part in visual interactions since they would be without their fuel. There are no rays that can be reflected by the pigment molecules; only blue rays that give their electrons a kick into a temporary new orbit and become consumed in the process. An orange (fruit) cannot be found with a blue torch. It can be found only with a beam containing orange rays (such as a white light).

Although blue is the only colour seen in the deep sea, an interesting twist emerges when its wavelengths are examined in detail. We casually assign the name 'blue' to a *range* of wavelengths, which is actually made up of various shades of blue. And the blue part of sunlight that remains from 200 to 1,000 metres, and the blue wavelengths seen by deep sea eyes, are *different*. Deep sea eyes in fact see the wavelengths of bioluminescence. So somewhere along the line there has been an interesting side-step in evolution. Rather than simply evolving bioluminescence with the same wavelengths as the sunlight blue, which could be seen easily, slightly different wavelengths evolved – bioluminescent blue. And eyes evolved to adjust – bioluminescence became more important to animal behaviour than sunlight below 200 metres depth. Maybe the evolution of bioluminescent blue happened because it was the most direct, energy-efficient option, given the foundation chemicals from which it evolved. Anyway, that's the situation. In the deep sea eyes see only bioluminescent blue.

It is likely that bioluminescence originally evolved in the deep sea for mating purposes – to signal to a potential mate, providing details either of one's whereabouts or of one's fitness. This would have ensured the retention of eyes in deep sea animals, in conjunction with their adjustment to bioluminescent blue. Then prey animals would have exploited this situation – they evolved flashing lights that confuse or blind their predators, buying them extra seconds to escape. All very conventional and conveniently explained.

'Oar-footed shrimps' or 'copepod' crustaceans (from the Greek, in reference to their broad, paddle-like swimming limbs) are tiny relatives of the familiar shrimps and crabs. Many of those that inhabit the deep sea are red, yet are eaten by visually guided fishes. But *how* does the predator fish find something that is red when the only available light is apparently blue?

From a deep sea submersible, a thermal imaging camera, through which only an animal's body heat (and no light) is imaged, is employed to detect the oar-footed shrimps swimming with their paddle-like limbs. A medium-sized fish approaches from the left of the image, clearly spots one of them, and accelerates directly towards it. 'Snap!' The fish now has the thermal image of an oar-footed shrimp in its stomach. One by one, the oar-footed shrimps fall from the water, picked out by the fish like flies by a chameleon. The fish sees them, swims towards them, and eats them. But *how* does it see them? From the facts outlined so far, this does not make any sense. Again, a red animal cannot be found under blue light. The only thing we can be sure of is that the oar-footed shrimps *are* seen and then eaten.

Maybe the best approach to this problem is to consider camouflage in the open sea in general. How do animals appear inconspicuous in open water, where they are potentially viewed from the front, back, sides, above and below? Well, there are five common tactics employed by animals in answer to this.

Transparency

Probably the most obvious solution to appearing inconspicuous in open water is to be the invisible man – to become completely trans-parent. But this is more easily said than done. By their nature, some parts of the body lend themselves well to evolving transparency, but others are more problematic.

Take an oar-footed shrimp. It has a hard exoskeleton or shell like that of an insect, and this conceivably could evolve to become trans-parent. Simply lose all pigments (including fluorescent ones), adjust the thicknesses of the internal layers so that they do not act as multilayer reflectors or liquid crystals, smooth out all layer surfaces to exclude dif-fraction gratings, and stay clear of particles in solution that could cause scattering. The same applies to the casings of the internal organs and vessels. But the contents of those organs and vessels are another story. They may lie beyond the control of the host animal. What happens when it swallows a meal, for instance? I remember when I found the transparent seed-shrimp because it had consumed a tiny opal, and the transparent crab because it had eaten a bioluminescent seed-shrimp. Both could not be more obvious.

Oar-footed shrimps have a light receptor used to detect light condi-tions. Like an image-forming eye, the receptor is dependent on light-sensitive pigment in its 'retina', which *cannot* be transparent – it must absorb light in order to convert it to electrical impulses. Light-sensitive pigment is usually dark red or black. In shallow waters, this will always be conspicuous.

But all is not lost. If *most* of the body can be made transparent, this will constitute an advantage. It gives a predator less to work with during a hunt. And then the dark red spot would not appear as an oar-footed shrimp but as something else, conforming to the rules of disruptive coloration.

Many animals do achieve near-total transparency. Crustacean larvae, such as those of crabs, live as plankton, swimming in open water. These are usually transparent, again visible only by their eyes. Some fishes also have totally transparent bodies except for their eyes. But the grand-fathers of transparency are those animals that have been transparent for probably the longest.

It's easy to be stung by a jellyfish – you don't see them coming. Jellyfish are among the least derived multicelled animals that exist – they have one of the longest evolutionary histories – and possess just two different tissue layers (one a protective skin, the other with a diges-tive function) separated by a gelatinous (watery) material. Such

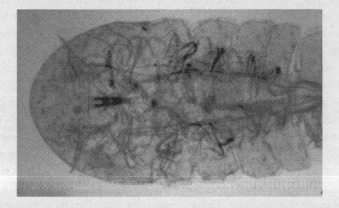

Figure 8.1 A transparent oar-footed shrimp (*Sapphirina* female – the males are iridescent). Light levels are adjusted so that the exoskeleton is just visible (internal organs and muscles are stained and so appear dark).

simplicity makes the evolutionary explanation of transparency in jelly-fish comparatively trivial. And in many jellyfish, all their layers *are* transparent. That serves them well for avoiding predators, but also for creeping up on prey. But occasionally jellyfish blow their invisible cover.

Although retaining a body made only of transparent materials, in some species their outer layer becomes so thin that it forms an optical thin film, like a fly's wing – it becomes iridescent. Similarly, in other species an outcropping of the skin concertinas for the purpose of rigid-ity, to provide a swimming 'oar', but at the same time it exposes a diffraction grating to sunlight. Again, another spectrum is dispersed into the water (the related comb-jellies are most guilty of this, although their iridescence may have evolved specifically *for* conspicuousness). But then we can see *all* jellyfish in the water eventually, given time to adjust our eyes, even those made of completely transparent materials. That's down to surface reflections.

Remember, from the Violet chapter, that thin film and multilayer reflectors work by refractive index differences. When two transparent materials with different refractive indices are pressed together, light will reflect from their boundary. Well, the two tissue layers of the jel-lyfish have refractive indices that are similar (around 1.5), but a little different to that of water (1.33). The bigger the difference, the stronger the reflection, and in this case just enough sunlight to be seen is reflected from the outer and inner surfaces of the jellyfish as they inter-face with the water. This effect becomes more serious to animals on land because then the difference in refractive indices becomes greater, since air has the lowest refractive index (1.00). Here the reflection from the surface of transparent materials becomes quite obvious – look at a bee's wings, for instance. Now we are comparing a refractive index of about 1.5 with 1.00. The bee does not strive for inconspicuousness, as one look at its body will tell you, but for some butterflies with trans-parent wings, surface reflections become a matter of life or death. Specialised 'anti-reflection' structures have evolved on the wing surfaces to reduce the reflections by a factor of ten. The anti-reflection struc-tures effectively cause the wing material and air to *gradually* merge into each other, without leaving a boundary open to strong reflections. Transparent butterflies achieve invisibility after all.

Figure 8.2   The highly reflective wings of a bumble bee (left), and the non-reflective wings of a hawk-moth (right). The wings of both insects are made of transparent materials.

Figure 8.3   An electron micrograph of the anti-reflective surface of a hawk-moth's wing. The wing is without scales, and these nodules break up the otherwise smooth surface and impart a gradual transition between the wing material (chitin) and surrounding air. Effectively, light 'recognises' no boundary from where to reflect. The white scale bar represents 1000 nanometres.

Then there is a further pitfall for animals with a transparency scheme – polarisation.

We know that a beam of light contains individual rays of either different wavelengths, as in sunlight, or a 'single' wavelength, as in bioluminescent light. The path of each ray can be traced to reveal its wave profile, involving peaks and troughs. But the wave profiles of all rays may not occupy the same *plane* within a beam. The rays all head in the same direction, but may oscillate in different planes. Consider a beam passing directly across this page. Some of its rays may oscillate from the top of the page to the bottom, which may be traced as a wave on the page, whereas others may oscillate in to and out of the paper.

As light passes through a transparent material, its polarisation properties may change. A beam of rays with different planes may enter the material, but only rays oscillating in a specific plane may make it through to the other side. Rays in all other planes may be absorbed, due to the orientation of the material's component parts. Just as a letter can enter a postbox in just one orientation, only rays of one plane could pass through the gaps between microscopic bars; others would hit the bars and be absorbed. So the light leaving a material such as a jellyfish's body may be different to that entering it – it may have become 'polarised'. This is how Polaroid sunglasses work – the lenses are filters that allow only one plane or 'polarisation' of light to pass between its elongated molecules ('bars'), arranged in parallel, and to reach the eyes. The quantity of light reaching the eye becomes greatly reduced – it appears less bright.

On land many animals, including bees, use the variation in polarisation planes in the sky to navigate. Sunlight itself is unpolarised, as are the portions of the sky near to the sun and directly opposite it (180° away). But the portion of the sky between these two regions (90° away) is partially polarised by the Tyndall scattering effect in the Earth's atmosphere. So with polarisation-sensitive vision the position of the sun in the sky can be known even on a completely cloudy day. The dung beetle even manages to roll its dung balls in a straight line at night using the polarisation properties of moonlight for orientation. And while I'm on the subject, the Vikings looked through crystals to detect

Unpolarised   Polarised

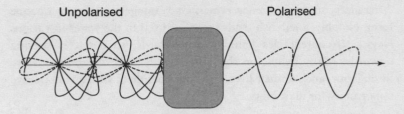

Figure 8.4   The possible change in polarisation as a beam of sunlight, travelling from
left to right, passes through a transparent object (shaded). In this case the
transmitted light is said to be 'plane polarised' since it oscillates within a fixed plane.
Alternatively, depending on the molecular properties of the object, the transmitted ray
may twist as it continues its path, again in the same direction.

polarised differences in the sky, and so navigate, when the sun was not
visible.

The transparent wings of some butterflies selectively polarise light,
which they use as a mating signal – they have polarisation-sensitive
vision to complement. But there are marine animals, such as squid, that
also detect differences in polarisation. Squid, for instance, have a retina
without cone and rod cells, but with long light-detecting molecules
that are arranged in groups lying either parallel or at right angles to
each other. This organisation is ideal for sensing polarisation differences – light rays with the same axis as these molecules are detected
best.

This is bad news for a transparent prey animal such as a fish or crab
larva. If a squid, as it swims through the water, suddenly detects a
change in the polarisation of the sea's light field, from near-unpolarised
to strongly polarised, its alarm bells will ring. It will recognise that to
become polarised, the light must have passed through a prey animal's
body. The squid simply swims in the direction of the polarised beam,
and eventually meets the polarisation filter – its food.

So once more a variation emerges on the concept of 'seeing is seeing
colours', as assigned in the Ultraviolet chapter. Not only is brightness
another aspect of seeing, but now polarisation appears important too.
Squid can detect a transparent fish using *only* its polarising effect.

Nevertheless, not all the predators of a jellyfish detect polarisation differences, so transparency remains advantageous over an opaque form. Notably, a jellyfish tends to orient itself in the water in various ways – from tentacles hanging downwards, to tentacles waving upwards. Many fishes, on the other hand, hold the *same* posture throughout much of their lives. This provides them with an alternative opportunity for invisibility.

Countershading

Taking an overview, natural history cameras scan a school of squid as they swim through the open ocean, near the surface where sunlight best fuels their vision. They are conspicuous, signalling to each other using their colour-change language. Suddenly one disappears from the middle of the camera's view, even though the wide-angle lens is employed. Then another vanishes, followed closely by another. The squid are not suddenly switching to camouflage colours, because others in the vicinity continue to be conspicuous (it's all or nothing for camouflage within a school of squid). Something is taking them. But nothing else is observed in the water, until the surface is broken, that is.

Even from beneath the surface, the imposing black marlin is noticeable as it leaps from the water and exhibits itself in the air. As it exits the water it leaves behind its camouflage – a fish out of water will always be conspicuous. But in the very different light conditions of the sea, the marlin is all but invisible.

In the surface waters, sunlight continues as a strong beam, on a course directly from the sun. Looking upwards, towards the surface from below, the background appears bright. Looking down from the surface, on the other hand, the background becomes the dark blue of the deep (Tyndall scattering from below results in some blue light reflected back towards the surface). To have a background that appears different depending on the observer's position places a new constraint on camouflage in mid-water, and the evolutionary response can be found in most shallow-water fishes of the open ocean – countershading.

Countershading is simply camouflage ideal for a mid-water

Figure 8.5 A countershaded black marlin – conspicuous out of water (left), yet all but invisible under water (except for the white inside of its mouth and data tag in this case) (right). Photographs by marlin biologist Julian Pepperell (left) and Greg Edwards (right) taken without a flash (a flash would change everything by adding light to which the fish is not adapted).

lifestyle, within sunlit depths. To appear inconspicuous against a dark background, an animal must be dark coloured too. But to achieve invisibility against a bright, white background, the animal must be white. Fortunately, a predator does not see both backgrounds at the same time, and this provides fishes with an evolutionary option, all because they remain horizontal in the water. In this posture it is only their upper halves that are viewed against the dark background, and so these are dark coloured to match. And their lower halves are only ever viewed against the bright, white background, so they too become white. From the side, the fish is divided along the centre of its body, from head to tail, into a dark upper half and white lower half. This 'countershading' strategy must work well, since it has evolved independently many times in fishes. Not surprisingly, then, countershaded fishes appear near-invisible in the water but conspicuous out of water, when only their background changes.

Many large sharks are countershaded, with white under surfaces and dark blue, grey or brown upper surfaces. Most other fishes living in the upper part of the ocean also reveal similar upper surfaces, but the white of their undersides takes a different form – silver.

Mirrors

Silver is the effect perceived when white light is concentrated into a strong directional beam, as opposed to scattering, where the sun's rays are spread equally into all directions. So walking through an environment we would see nothing of a silver reflection until our eyes intercept the reflected beam, then the silver suddenly appears. It is the brightness (the perceived intensity) that separates silver from white. Their broad ranges of wavelengths are the same.

While painting *The Blinding of Samson* in 1636, Rembrandt used streaks of white paint on his soldiers' armour to provide the illusion of metal. But bare-metal armour appears silver. Unfortunately the metallic effect cannot be fully reproduced using pigments, which have the same effect when viewing the canvas from all directions. But like Monet's attempt to reproduce the metallic-like green flash from pheasants' heads, Rembrandt achieved some success by maximising the colour and *brightness* contrasts. The metal armour immediately surrounding the white reflection, along with the soldier's background, is depicted darkly, out of the domain of any light source. This is the best an artist can do.

Although solid metals can have the type of surface to provide a silver reflection, in nature silver can result only from an optical device made of transparent materials – it is a structural colour. The light beam is so much more intense from a structural colour compared to a pigment – perhaps in some instances ten times more. But as I have hinted throughout this book, intensity is the actual property of a light beam and brightness is how we perceive it. Brightness is to intensity as colour is to wavelength.

When it comes to intensity, the eye is a logarithmic detector – it does not convert intensity *directly* to brightness. So if the structural colour in the real world is ten times more *intense* than the pigment on the artist's canvas, then it will appear only twice as *bright* to us. The reproduction of structural colours in art is not a hopeless case after all. And by using black or another dark colour to surround the white, silver-imitating pigment, then the brightness contrast helps the cause. In *The Blinding of Samson*, Rembrandt was careful to keep his soldiers in a dark cavern where a narrow beam of sunlight could burst through the entrance and strike their armour but not their background.

So if silver in nature is structural, then what kind of structure is involved? To answer this question we must return to the subject of multilayer reflectors, as covered in the Violet chapter. But after learning the ins and outs of these optical devices already, the explanation for silver can be kept short.

The multilayer reflector responsible for the violet colour of the blue crow butterfly was simple, in that all the layers of chitin in its wing

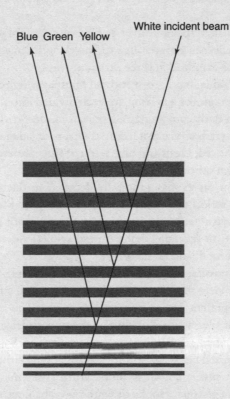

Figure 8.6 A diagrammatic section through a silver (broadband multilayer) reflector, showing how different colours are reflected from different layers, but all into the same direction, so reforming white or rather silver.

scales were of the same thickness. And thickness is partly responsible for the wavelength or colour reflected. So the blue crow reflector appeared as *one* colour in a *single* direction.

Now what happens if the layers of the same material within a stack become *different* thicknesses? Simple – each layer reflects a *different* colour in a single direction. That is, in *one* direction, some layers will reflect blue rays, others green rays and others red rays, with layers for other colours in between. These rays add up to form . . . white. And since the rays are concentrated into a single direction, they appear bright. A bright form of white is *silver*. This type of reflector acts as a mirror.

The stack that appears silver is known as a 'broadband' reflector because it reflects a broad range of wavelengths in each direction. Its layers may be arranged in three different ways.

First, individual blue, green and red multilayer reflectors may simply overlap. The *scales* of a herring are each divided into thirds of different colours, one third containing red reflectors, one green and one blue. The scales overlap on the fish like tiles on a roof, so that at any position on the fish the red, green and blue parts of three different scales will lie on top of each other.

Second, the layers may gradually decrease in thickness from those suitable for reflecting red at the top of the pile. Moving down the stack, the layers become progressively thinner until they end with a layer suitable for reflecting violet (or ultraviolet). Silver beetles employ this type of broadband reflector.

The third possible arrangement is where the layers of the second type are shuffled. Here the layers suitable to reflect each colour in the spectrum are all present, just in no particular order. This reflector is found in the *skin* of silvery fishes, including the undersides of marlins.

Curiously, the study of silver in animals dates back to Robert Hooke, when he reported his 'microscopic' investigations of the silverfish insect ('the small silver colour'd bookworm') in his famous thesis *Micrographia* of 1665. Hooke drew the silverfish, including the microscopic scales that cover its body. He attributed its silver colour to these scales.

Hooke's explanation of how thin films were responsible for the silver

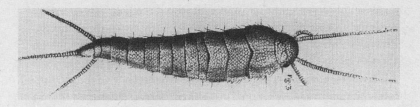

Figure 8.7   Drawing of a silverfish insect reproduced from Hooke's *Micrographia* (1665).

colour could have been no more than a guess, since the films or multiple layers are visible only with a modern electron microscope. But was Hooke's guess lucky or educated? Like Newton soon after (in his book *Optiks* of 1704), he too noted that if glass is blown into 'exceedingly thin' layers and 'laid in a heap together' these have the same appearance as a pearl (I wonder, was Newton influenced by Hooke's observations, or was his venture into glassworks a coincidence?). A pearl does employ broadband reflectors, although in a complex way. And since light was not known to contain the spectrum of colours in Hooke's time, did he properly *understand* what he was insinuating? If Hooke's guess was educated, he may have pipped Newton at the post in assigning the first structurally coloured animal. I would not have suggested this in Newton's time, since the two were mortal enemies and Newton was the political heavyweight, and even now will leave a question mark next to Hooke's silverfish revelation. In any case, with my physics colleagues David McKenzie and Maryanne Large of Sydney University, I examined a silverfish and found the silver reflector not within the scales but within the exoskeleton beneath them. But that's a minor point.

Returning to fishes, their silver colour often serves a purpose unexpected for a bright light – camouflage. In the sea, the light field becomes increasingly even the deeper one travels, although always retaining a downward bias. That is, the directional beam of sunlight is broken up to illuminate marine animals from all directions to some extent. And in such light conditions the mirrored sides of a fish will always reflect towards an observer light that would be seen if the fish were not there. A predator looking directly at the fish from below sees only a reflection of downwards-travelling sunlight from the fish's sides, but not an image of the fish itself. Effectively the fish disappears from the sea.

All is not totally idyllic for the mirrored fishes. They suffer the same shortcomings as transparent animals – their mirrors alter the polarisation of sunlight. Unpolarised light strikes them; polarised light is reflected. So a predator with polarisation-sensitive vision, such as a squid, would detect the silvery fish after all. But camouflage in colour is better than nothing. And not all predators of silvery fishes have polarisation-sensitive vision.

This form of camouflage – reflecting the environment in one's body –

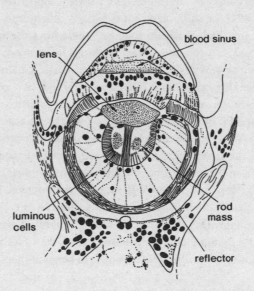

Figure 8.8   Section of a photophore, or bioluminescent organ, of the (euphausid) shrimp *Nyctiphanes norwegica*. Light is projected towards the top of the page. Adapted from an 1888 drawing by R. Vallentin and J.T. Cunningham.

can be found also in some butterfly chrysalises of forest canopies. Here leaves break up sunlight so that it also strikes the animal equally from all directions. The chrysalises are silver or gold (gold is silver without blue) but appear as only the precise shade of green of the leaves that surround them. In direct sunlight, on the other hand, they look like small, polished nuggets of silver or gold. They become so shiny you can see yourself in them.

Sometimes mirrors occur within chromatophore cells, where the mirror effect can be turned on or off just like the colour effect of their pigmented counterparts. These are employed by some octopuses, including *Octopus vulgaris*, to reflect any colours in their surroundings that are not covered by their range of pigmented chromatophores. In particular, blues and greens must be reflected, since these octopuses have only yellow and black pigmented chromatophores (and sometimes orange and red).

Mirrors are also found in the eyes of some animals, to focus light on to the retina in the manner of a Newtonian telescope, and within bioluminescent organs, to direct harmful intensities of light away from the body. Many shrimps are bioluminescent, and often have a lens and a mirror in their bioluminescent organs, or photophores. These photophores look similar to those of fire-flies, with a bowl-shaped reflector lying behind their luminous cells, but above those cells is a lens to focus the bioluminescent light into a beam. And then the reflector of shrimp photophores is much more efficient than the simple scattering system of fire-flies – it is a concave mirror. This further directs light into a beam but also is very efficient at reflecting the bioluminescent light away from the vulnerable host tissues beneath.

Alternatively, an 'amphipod' crustacean (a small, shrimp-like animal) is known to reflect the bioluminescence produced in its head using a dish-shaped (concave) mirror on its antenna. The mirror concentrates some of the bioluminescent light radiating from the head into a narrow beam that probably draws up tiny, light-sensitive prey towards its mouth. But in complete contrast, beams of bioluminescence can also provide camouflage in the sea.

Bioluminescence

Post First World War military aircraft carried the advice 'best flown at night'. Although the planes were painted with camouflage colours, during the day they stood out as silhouettes. The problem lay in 'earth-shine' – there's not enough light reflecting from the Earth's surface to illuminate a plane's undercarriage. The pale undercarriage of a military plane does not receive a light source bright enough to fuel its pigments – sunlight falls mainly on the top and side of the plane. Countershading is not an ideal strategy in the air, although is better than nothing. For instance, a *white* gull appears as a *black* shape as it glides directly overhead, although its underside does appear pale and less conspicuous when the gull banks.

Then, in the 1930s, an interesting question was posed in a defence research lab. 'In the absence of a suitable light source for camouflage colours, why not make one's own light?' The implications involved fixing lamps to the undercarriage of a plane, and shining down to Earth the same intensity of white light that falls upon the plane from above. A photocell on top of the plane would sense the downward-travelling light, and adjust the intensity of the plane's lamps to match. Then anyone looking up towards the plane would see only apparently natural sunlight beaming down on to them.

Lights were fitted to the undercarriages of test planes, and worked according to plan, but there was teething trouble all the same. The lamps jutted out from the underside of the plane's body and wings, causing significant drag. And then a more serious concern. The camouflage effect was achieved only when the plane lay directly overhead, but by this time it could be heard anyway. The lamps, it seemed, solved one problem but fashioned another. But from a different military drawing board sprang a compromise solution.

'Why not engage the lamp camouflage system when the planes were too far away to be heard?' was the riposte. As a result of the US Navy's 'Project Yehudi', in 1942, TBM-3D Avenger torpedo-bombers, and then B-24 Liberator bombers, were flying over the Atlantic with lamps fitted to their *leading edges* – on the nose of the body and front edges of the wings and tail. The photocells were moved suitably to the trailing edges.

The planes were now camouflaged when they appeared on the horizon. In fact they could be seen from only two miles away, as opposed to twelve miles away without the lights. Then Project Yehudi was dealt a fatal blow. Newly invented Radar became fine-tuned and deployed.

When radio waves are sent prospecting the skies, a plane on the horizon will bat them back to their transmitter, making their presence known. Needless to say, telescopes for spotting planes quickly went out of fashion. But whether known or not by the architects of Project Yehudi, deep sea fishes carried the concept into the future, and had in fact employed it long before.

Peter Herring, of the Southampton Oceanography Centre (England), is an authority on marine bioluminescence. He has spent many hours in deep sea submersibles, watching the firework displays of bioluminescent flashes as the submersible's movement lights the touchpaper. As some squid illuminate their whole bodies in blue, a jellyfish sets off its spiralling lights, spinning backwards and forwards like a winning fruit machine. Peter's videos of this are mesmerising – he is great value for a seminar. Other squids send a trail of flashing lights down a long, dangling tentacle, while the bioluminescence of other jellyfishes additionally fuels blue pigments, blue-stimulated fluorescent pigments and structural reflectors in their bodies. Four colour factors in one go. And then there is the hatchet fish.

The hatchet fish has a flattened body with silver sides. This body shape is a good strategy to reduce the size of one's silhouette from below. The hatchet fish swims in open water at between 200 and 1,000 metres depth, where sunlight is blue and mostly streams downwards, carving out a silhouette of the fish when viewed from underneath. It suffers the same limitations as airborne animals – the hatchet fish's predators may pass directly below, making it potentially obvious in their eyes. As a predator swims, its upward-directed eyes detect sunlight that would theoretically become broken by the passage of a hatchet fish above. But the hatchet fish is bioluminescent too.

In Yehudi fashion, the hatchet fish shines its bioluminescent lights downwards – mainly directly downwards. Its eyes detect the brightness of the light falling on its head, like the photocells of 1930s military planes. And from below, a hatchet fish is just not visible, at least to us.

Figure 8.9   A hatchet fish from mid-deep water. From the side in bright light (top picture). From below in darkness and without a camera flash (bottom picture) – the spots of light are glowing (blue) bioluminescent organs. Photographs by Peter Herring, courtesy of Image Quest Marine.

Its lights seemingly provide a perfect match to the adjacent sunlight. Well, near perfect.

Bioluminescence carries a slight handicap in this case. Although essentially blue, it tails off just into the yellow part of the spectrum. There is some, albeit minimal, yellow in marine bioluminescence.

Between 200 and 1,000 metres depth, in the mid-deep sea, sunlight is *strictly* blue. It contains *no* yellow. And here lies a selection pressure for the hatchet fish's predators.

Fishes that feed on hatchet fish have evolved yellow lenses in their eyes that act as filters. Solely yellow light is allowed to reach their retinas and be seen. This can only be a strategy that permits the detection of hatchet fish and other prey animals with bioluminescence camouflage. Sunlight is not seen at all by these eyes, so the minimal amounts of yellow in marine bioluminescence stand out. Clearly fishes with yellow lenses must be extremely sensitive to low light intensities. But the yellow-lens strategy has to be effective – it has evolved over ten times independently in predatory fishes (over ten different types of yellow pigment are found in the lenses of these fishes).

Below 1,000 metres, where blue sunlight is absent, both bioluminescent camouflage and yellow lenses are obsolete. Here, in the *deep* sea, another approach to camouflage dominates.

Black and red

In the fourth century BC, Aristotle and other Greek philosopher-naturalists took an interest in nature. They made many observations about plants and animals, including those of the seashore. But the founder of the *science* of oceanography and marine biology is generally credited as Edward Forbes. In the early nineteenth century, Forbes, a British naturalist, laid the foundations for the study of marine life at the ocean surface and on the continental shelf – the first 200 metres of sea floor. Beyond the gently sloping continental shelf the sea floor suddenly plummets to 2,000 metres down a greater slope. Forbes was without equipment to sample beyond the continental shelf, and was faced with plenty to occupy his time nearer the shore in any case, but that does not excuse his 'azoic' theory. This followed that life did not exist on sea beds over 600 metres deep.

The British *Challenger* Expedition, from 1872 to 1876, well and truly sank the azoic theory. This is considered the first 'modern' oceanographic expedition because it standardised methods to collect data and investigated aspects of not only biology but also physics, chemistry and geology. The *Challenger* was the first oceanographic research vessel to circumnavigate the world, covering all the major oceans except the Arctic, and the final report from its journey occupied over fifty volumes, which are still useful today. They established the existence of life at extreme depths. Then, just before the turn of the twentieth century, the German *Valdivia* expedition set sail with equipment to focus specifically on the deep sea, and more than seconded the *Challenger*'s declarations of life at depth.

This work opened the doors for marine biological laboratories all over the world. Among the first was the Statzione Zoologica de Napoli, established in Italy in 1872. A year later Luis Agassiz founded the Anderson school in Penikese, USA, which later became the Marine Biological Laboratory at Wood's Hole. In 1888 the Marine Biological Laboratory in Plymouth, England, began, and in 1906 the Oceanographic Museum and Aquarium in Monaco first opened its doors. Prince Albert I of Monaco was an avid marine biologist, conducting numerous research expeditions. Likewise, King Oscar II of Sweden supported the deep sea cause, while neighbours Norway and Denmark contributed famous expeditions of their own (the *Michael Sars* expedition and the *Dana I* and *II* expeditions respectively). By the beginning of the twentieth century, deep sea biology was in full flow. Of course the subject stepped up a considerable notch with the advent of manned submersibles in 1934, although even today they are not likely to meet each other. Still we have yet to scratch the surface of deep sea biology – the deep sea remains the least known environment on Earth.

From what we do know, deep sea animals rarely encounter a meal, so when they do, they must make the most of it. This alone can explain some of the bizarre forms exhibited by deep sea animals. Indeed, some appear like a swimming mouth. And mouths that at first appear normal will suddenly swing open on hinges to engulf large prey with inconvenient shapes. The extreme pressure does not encourage life to conform to shallow water rules either, and also explains why the pressure-adapted, soft-bodied deep sea animals transform to very different shapes when they are brought to the surface. This was one stimulus to send down manned submersibles originally. But the colours of those specimens hauled to the surface by all those deep sea research vessels, beginning with the *Challenger*, *did* remain natural and provided us with the deep sea's answer to camouflage.

Beyond 1,000 metres depth exists a common, energy-efficient option for inconspicuousness. There is no light to fuel countershading or mirrors, and no downwards-travelling sunlight to justify bioluminescent camouflage. Transparency remains a possibility, and is evident in the deep sea, but a far more widespread and often simpler solution is to be coloured completely black or red.

The deep sea environment is the largest and most stable on the planet. Temperature and pressure are constant all year round – it is always cold. A submersible provides a good indicator of this. As one submerges, the temperature change causes water vapour within the air in the vehicle to condense on the ceiling of the vehicle. The water proceeds to drip on to the head of the pilot (and according to Peter Herring, the sensation of water dripping on the head at helpless depths is rather alarming). In fact the only unstable characteristic of the deep sea is light.

Although the deep sea appears generally black, with no cycles of light and dark, the visual monotony is broken by bioluminescent flashes from near and afar, as described at the beginning of this chapter.

Deep sea fishes tend to be weak and swim slowly, a trait echoed in their eyes. Contrary to original ideas of degenerative eyes, the eyes of deep sea fishes are extremely well adapted to detecting flashes. And all they have to see in the deep sea *are* flashes of bioluminescence. Also, they see the flashes with great precision, albeit only to a distance of a few tens of metres. But then their weak bodies could not reach an object beyond this distance in reasonable time anyway – the vision of deep sea fishes serves ideally their behaviour too.

Fish have eyes that are similar to our own. Up until now I have talked about only the cone cells in our retina. But there exists, alongside the cones, another type of light-receptive cell – the rods. Rod cells are sensitive to low light levels and are employed by us at night. They see only in black and white.

Shallow-water fish have cone and rod cells in their retinas too. The cone cells, of four types, are adapted to see the full spectrum of colours (including ultraviolet) that exist on coral reefs, for instance. But fishes that are active during dawn and dusk have more rods than cones. And the range of cone cells becomes reduced as the environment becomes deeper – at 100 metres depth fishes have only the blue and green cone types. Finally, beyond 200 metres, the deep sea fishes have only rod cells. It is these rods that are sensitive to blue, bioluminescent wavelengths, allowing the fish to see in monotone, or black and 'blue'.

As I mentioned, other than some transparent species, most of the animals brought to the decks of research vessels during their history of deep sea sampling were coloured red or black. Although black can appear most conspicuous against a pigmented background, through the colour and brightness contrast it provides, against the darkness of the deep sea it is invisible. And lit by only blue, bioluminescent torches, which at the most stretch into the yellow, red is completely invisible too. Maybe in some animal groups red pigment is less demanding of evolution than black pigment.

As the *Challenger*'s deep-sea nets were opened, many red-coloured animals were sent sprawling across the deck. Red crabs, shrimps, fishes, jellyfishes, octopuses, squids, worms, sea stars . . . a whole diversity in red. Red *pigments* were always to blame – the colour of these animals has faded away through the years of storage in alcohol, as only the hues of pigments can. Fortunately, the nineteenth-century publications from deep-sea expeditions, such as the *Challenger* volumes, afforded the luxury of hand-coloured plates, based on original notes of colour made on the ships' decks. Again, scientists at this time were the sports stars of today, and many walked on red carpets. Fortunately, their works preserve the colour of the deep sea fauna. We are told that the leading lights of this chapter, the deep sea oar-footed shrimps, were red too.

An opportune intermission

Joining a submersible at the beginning of its daytime descent, a camera confirms all that has been explained so far in this chapter. Marine animals are certainly difficult to find under their natural lighting, but become quite obvious once headlights are employed, to which the animals are not adapted. But, while still at only thirty metres depth, one animal becomes extremely obvious, although through no fault of its colour.

'Crack!' Even within the submersible, the snap of a snapping shrimp is quite obvious. Indeed, it frequently disturbs naval and scientific sonar, such is its audible intensity. The loud noise is made, of course, by a small animal – the snapping shrimp is just a few centimetres long – but one of its claws is large, appearing disproportionate with the rest of its body, and it is this that snaps.

evokes responses throughout the environment. Everything that can sets off its blinding-flash reaction in a frenzied attempt to dazzle or confuse an assumed nearby threat. And there is hope that, at the same time, the flashes may alert the predators of *this* threat – a flash of blue light in the dark is attention-grabbing. It attracts animals from as far as their eyes can see.

NanoCam is facing out into the environment, but it has become distracted by movement to its sides. The red pigment molecules lying next to it within the oar-footed shrimp's shell are set in motion. Blue light strikes them, they grab the energy of the rays, their electrons jump about, and the oar-footed shrimp warms up – the energy of the blue light is quickly conveyed towards the cold sea as heat. The red pigments are working. They have absorbed the blue rays and dumped their energy into the ocean as heat. All accords to assumption.

The pigments work hard with each new and renewed bioluminescent flash, followed by a gradual reduction in electron-jumping as a flash fades. The oar-footed shrimp absorbs the blue bioluminescent rays and so the blue-sensitive eyes around them cannot detect it. The pigment molecules complete their cycle again and again, just as expected, but then suddenly there is something unforeseen. The pigment *reflects* some rays – *red* rays.

Red rays strike the oar-footed shrimp's molecules and of course bounce from them. The molecules are red pigments, and that's what pigments do – reflect red, and only red, rays.

The reflections become stronger and stronger as more and more red rays bounce from the oar-footed shrimp's pigments. Soon the brightness control of the NanoCam monitor is turned down, such is the intensity of the red light. But what red light? Red light should not exist in the deep sea. Then suddenly all becomes clear.

Focusing back on the environment, NanoCam is found to be within a red beam of light. The red increases in brightness more and more until suddenly the red light source is imaged, along with a red reflection from the shiny surface of long needle-like teeth. The teeth move apart, until . . . snap! Darkness. There is no more light on the monitor.

The oar-footed shrimp has succumbed to a dragon of the deep sea, or more precisely a dragonfish. The fairly small, elongated bodies of dragonfish are designed only for oar-footed shrimp predation, and are indeed a stealthy black to avoid larger predators themselves. But in addition to *blue* bioluminescent organs, they also possess *red* ones, and these breathe effective fire. Amazingly, in a world of blue, the dragonfish has evolved not only red headlights, but also the ability to see in them – its eyes are sensitive to red too. Dragonfish have their own, private wavelengths. In addition to lighting up their prey, maybe they also communicate to each other using them.

The dragonfish, with their red lights and vision, evolved after the red oar-footed shrimps. Then the situation became hopeless for the up until then visually adapted deep-sea oar-footed shrimps. Now their predators are illuminating them and they have absolutely no idea it's happening. The only warning they have of dragonfish is the movement of water as their jaws are swung open. And by then it's all too late. One by one the oar-footed shrimps are effortlessly picked out of the water. However, probably we are witnessing a single moment within an on-going arms race where an evolutionary 'move' has just been made by one side, the dragonfish. The other side could make the next move, whereby the red pigments of the oar-footed shrimps evolve to be black.

The dragonfish is a recent discovery that mocks the established rules of deep-sea biology. Textbooks must be re-edited so that marine bioluminescence is no longer exclusively blue, and deep sea eyes are not always blue light detectors. And then there's the challenge for evolution – how red bioluminescence and red vision evolved simultaneously in the dragonfish, since one does not make any sense without the other in an otherwise blue world.

Researchers are considering the possibility that red visual pigments in dragonfish have their evolutionary origin in chlorophyll. Chlorophyll absorbs red light, and may cause the blue-sensitive rod cells in the retina to respond to red also. In theory, chlorophyll could derive from the bacteria of the *oar-footed-shrimp's* diet, and somehow make its way from the predator's stomach to its eyes. At least that would explain why dragonfish eat only red oar-footed shrimps – to maintain their red vision, in a very vicious circle. But then this would represent the first known use of chlorophyll in an animal. Biologists including Ron Douglas of City University, London, and Julian Partridge are currently searching the fish's anatomy for answers.

And the hunt is on for more dragonfish to study. So far, three species of dragonfish have been identified, but only one specimen has been found alive. According to Peter Herring, who captured it while working with Edie Widder of Harbour Branch Oceanographic Institution in Florida, 'We shone a red torch at it and it flashed back. But it didn't know what we were saying – and we didn't know what it said.' To repeat the words of Charles Darwin, 'He who understands baboon would do more towards metaphysics than Locke' (Notebook M, 16 August 1838).

A tonic for Darwin

I admit, the final colour *factory* introduced in this chapter (shrimpo-luminescence) was added only to fulfil my promise of seven colour factories on nature's palette. The red bioluminescence of the deep sea is rather more interesting to the biologist, and indeed holds another staunch message for Darwin. Dragonfish are hunted by large deep-sea predatory fish, which possess eyes that cannot see their blindingly obvious prey as they swim in front of them. Their prey send strong beams of light through the water, sometimes even clobbering the large predators on their heads, yet these predators remain totally unaware. The eyes of these large predatory fishes do not serve them well – *they are not perfect* by any means. Darwin, again, was wrong to worry about the eye. If he had not promoted it to 'an organ of extreme perfection', he would have had nothing to explain. His theory of evolution *was* strong all along. The *imperfections* in the eye act as selection pressures *for* evolution – the dragonfish literally get away with murder.

So that's it for the animal spectrum. I have covered all colours seen by animals and all the microscopic factories employed to make them. Now I can relax, turn on my HiFi using its remote control – fitted with an *infra-red* beam – and pour myself a . . . wait a minute, why did my goldfish suddenly turn in its tank to directly face me, or rather the remote control, as I pressed a button? Maybe this book, and the animal spectrum, is to be continued . . .

Conclusion for Evolution

When the world woke up to vision

The Hubble telescope continues to peer deeper and deeper into space. At the last count, the stars in its images were more than 75,000 trillion miles from Earth. This means that the light we receive from these stars was generated more than 13 billion years ago, taking us to within a stone's throw of cosmology's Big Bang (the origin of the universe, around 14 billion years ago). To put this into perspective, the light from Mars, around 125 million miles away, takes ten minutes to reach us. And again closer to home, our sun and solar system was formed 4.6 billion years ago. Sunlight has illuminated the surfaces of Mars and the Earth ever since.

This book is the second in a trilogy on the link between light and evolution, in particular the Cambrian explosion or evolution's Big Bang. This was the event, 540 million years ago, where animals from all major categories, or phyla, first evolved their hard parts, and in quite spectacular fashion. This book serves to reveal the sophistication and diversity of the visual world, and demonstrates how important colour and sight are to animal behaviour *today*. The first book in this trilogy, *In the Blink of an Eye*, considered the first eye to appear on

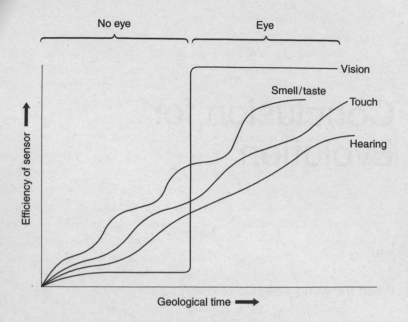

Figure 9.1   Graph showing the very approximate evolution of receptors for different senses throughout geological time. First published in *In the Blink of an Eye*.

Earth. I have reproduced a figure from that book above. It is the simplest, crudest diagram in the whole book, but one I consider the most important for the message it holds.

The figure demonstrates that vision is the only sense that can appear in the history of life within a very short period of geological time. Other senses, such as smell, touch and hearing, staggered into the history of life through gradually improving the efficiency of their sensors. They could not have unleashed a bombshell for animal life on Earth in general; the type of bombshell that could cause that life to explode into new body forms with hard parts. Vision, on the other hand, is a sense that exists either efficiently or not at all.

'Vision' demands an *image-forming* eye. This is not a minor detail, nor is it one simple to comprehend, and has been misinterpreted on numerous occasions. A light receptor that detects only light intensity, or the direction of sunlight, is *not* an eye. It will help enormously if only a light detector capable of image-formation is termed an 'eye', and other detectors of light simply as light receptors. A light receptor does not provide the capacity for vision, or make any of the chapters in this book viable. Animals cannot distinguish each other using such a detector, and so it can be considered perhaps 1 per cent efficient as a sensor for light. In stark contrast, an eye that grabs an image, along with a brain to interpret it, is close to 100 per cent. That's what the graph above attempts to illustrate, and that is the concept that must be grasped before the *Light Switch Theory* can be valued. As soon as a lens evolves within a light receptor, an image is formed, and the leap from 1 to 100 per cent efficiency for light perception becomes possible. The lens, along with a brain to process visual images, in turn makes viable the *interaction* of animals using light – the stimulus that has existed before animals themselves. Of course, the role played by other senses in animal behaviour today should not be underestimated, but as the graph above illustrates, they could not play such an important role in driving an *explosive* evolutionary event.

Another noticeable character of this 'evolution of receptors' graph is the shape of the 'Vision' line *after* the leap to an eye. It plateaus, implying stability of the general food-web structure. Once vision evolved for the first time, it did not go away. What we see today in terms of visual adaptation, as exemplified in this book, is what we would have seen soon after the first eyes evolved and thereafter (which conforms to the reasoning of other biologists, that animals function today as they did in the past – a concept we can take back as far as the Cambrian). So the greater the extent of visual adaptation today, the higher the importance we can place on the transition to vision in the Cambrian period. This book reveals not just a sophisticated diversity of colour in animals, but also whole bodies and costly behaviours designed specifically for visual adaptation. So it further fuels the *Light Switch Theory*, by demonstrating the gulf between body forms that are adapted to and maladapted to vision. In which case, only a *massive* evolutionary event could promptly bridge the gap once eyes had appeared. The shorter that event, the bigger the evolutionary bang. It is most likely that this forecasted event *was* the Cambrian explosion.

The first eyes belonged to a trilobite, some 540 million years ago. We know from the fossil record that just after the first eye evolved, life exploded. The Cambrian explosion or Big Bang of evolution took place, where animals from all major categories, or phyla, changed from soft-bodied worms to a variety of forms containing hard parts. Hard parts can mean visual adaptation, in terms of warning defences and fleet-footed appearances. And then strong swimming limbs are required to endure the chase by a visually guided trilobite.

When the adaptation of animals to the senses is considered over geological time, a relationship with the evolution of vision emerges. On a graph illustrating body and behavioural adaptations to the senses for *all* animals on Earth, the line for visual adaptation closely matches the line revealed for vision when the evolution of the senses is considered. The evolution of the first eyes on Earth sliced the history of animal life

Figure 9.2   Cambropallas from the early Cambrian period. This fossil trilobite, around the size of a small plate and 540 million years old, represents one of the first animals with eyes (here the eye on the left casts a shadow).

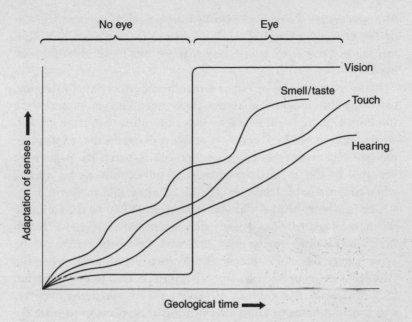

Figure 9.3 Graph showing the adaptation of animals to the major senses through geological time, including evolution of body forms and behaviour. The adaptation to senses other than vision is substantial also, but is a long-drawn-out process through geological time.

in two, into 'pre-vision' and 'vision' phases. Vision is the only major sense that could wield such a monumental sword. That all animals would respond equally dramatically could perhaps be expected.

Today we live in a visual world. There would be no such animals as stick insects, chameleons or birds of paradise – not to mention the cast of this book – if we did not.

Darwin's dilemma

For a trilogy of books that deals with light and evolution, it was always necessary to deal with the problems of Darwin himself on this subject.

The eye does stand out so awkwardly from the otherwise unruffled text of *The Origin*. So how has *this* book helped Darwin's cause? Well, it introduces Darwin to a discovery mapped out in another 'famous' book.

In the archives of the oldest, or rather the oldest continually running, university lies a seven-volume monograph called *Kitāb al-Manāzir* – the *Book on Optics*. The fragile grey pages are filled with Arabic handwriting interrupted by accurate ray-tracing diagrams that explain the paths of light rays as they encounter optical structures. The pages were prepared by Ibn al-Haytham in the eleventh century, as he worked while confined to his home near the Al-Azhar mosque in Cairo.

The University of Al-Azhar was established in 970 by the Fatimids, the political and religious dynasty that dominated an empire in North Africa at that time, forging close links with the nearby mosque of the same name. Ibn al-Haytham, also known as Alhazen, from the Latinised version of his first name 'al-Hasan', moved to this area from his birth town of Basra in Iraq. Al-Hakim, the Fatimid leader in Egypt, had invited Alhazen to Cairo after hearing of his plans to regulate the flow of water down the Nile.

The turn of the first millennium was celebrated with science in the Arab world. Intense research commenced in mathematics, physics and astronomy. Curved surfaces were measured, new geometrical methods invented, and the method of integral sums rediscovered. One hundred and fifty years later the first rigorous geometrical theory of lenses was developed. How the eye forms images on the retina became fully understood. And this is where Alhazen entered the scientific pursuit.

Alhazen headed a team of engineers in Egypt, but after travelling the length of the Nile he realised that his project to regulate the flow of water with large constructions (a kind of early Aswan dam) could not work. After his demotion to an administrative post he grew to distrust the ruthless Al-Hakim, and even considered himself in danger. As Alhazen saw it, he had just one option if he was going to survive the Al-Hakim reign – to pretend to be mad. This did indeed save his skin, albeit at the expense of confinement to his house in Cairo. But Alhazen turned this custody into an opportunity to write mathematical texts, including his *Book on Optics*. He was, nevertheless, compelled to wait until Al-Hakim died in 1021 before declaring his sanity and scientific genius to his peers.

Although the optics of lenses had been already figured out by Alhazen's time, a debate raged over how vision worked. On the one side were the 'extramissionists', who believed a visual ray was sent out by the eye, prospecting for objects. On the other side were the 'intromissionists', who proposed that objects send 'forms' of themselves towards the eye. Alhazen added his own idea to the pot – that light rays emanate towards the eye from every point of a visible object. He was on the right track, of course, but he went further and proposed another integral ingredient for vision.

Alhazen's *Book on Optics* reveals detailed experiments that led to a conclusion fundamental to the science of vision even today. His revolutionary finding was simple, yet it changed the way vision would be considered thereafter. Mathematical predictions of how a lens would act were as solid as Alhazen's experiments, but sometimes what Alhazen *saw* was a little different. It struck him that there were in fact *two* parts to vision – first a theory of light that included geometric and physical optics, and second the physiology of the eye and *psychology of perception*. There was light emanating from objects, and there was the interpretation of that light in our eye and brain. The two were different. The first could be calculated mathematically, the second could not.

Alhazen's ideas were translated to Latin, Italian and Hebrew, and lay at the foundations of scientific work in the centuries to follow. Roger Bacon, Frederick of Fribourg, Kepler, Snell, Descartes and Huygens were among the many prestigious names to be influenced. Unfortunately, one notable omission from this list is Charles Darwin.

Well, Darwin was never concerned with the fields of mathematics and optics, but, as it happens, he could have profited from reading Alhazen's *Book of Optics*. Had Darwin known of Alhazen's simple concept that vision involves the combination of light leaving objects *and* its interpretation in our brain, he may never have composed the 'Organs of Extreme Complication and Perfection' section of *The Origin*.

Darwin was in awe of the eye's optical system. He wrote of 'its inimitable contrivances for adjusting the focus to different distances, for admitting different amounts of light, and for the correction of spherical and chromatic aberration'. He continued to write as if the eye is our contact with the *real* world, one of our central means of reacting to and with our environment. Darwin subconsciously wove into his paragraphs a concept that the eye was so sophisticated that it revealed exactly what was out there before us. But that was the eye, not vision. The eye is the easy bit. Darwin failed to involve 'vision' in his argument, and it is vision that reveals our environment to us. The only effect that light has on vision is to change the shapes of molecules in the cells of the eye's retina. Nerve cells play their considerable part from there. Vision as a whole is an *unfaithful* sense.

Darwin had considered only one half of Alhazen's theory of sight – the eye and its associated optics. He had overlooked the physiology of the eye and the psychology of perception.

The eye *is* rather perfect and for that matter impressive when only its optics are considered. The *interpretation* of the images formed, on the other hand, is another story – a much more complex and exacting task. As this book reveals, the accurate interpretation of all conceivable images in nature by a processing unit that can fit within the spatial confines of a body is just not possible. Instead, what we find in nature is a compromise. The brain processes visual images in the most practical and appropriate way for the needs of its host. This method is *not* perfect.

Now we can consider that all animals with eyes, which include over 95 per cent of all multi-celled animals on Earth, live within a virtual reality world. What we see in front of us may not really be there. To begin with, there are no coloured objects in the world – the brain adds its own colour to objects, as established in each chapter of this book. And then we may see *some* objects, but not all of them. Indeed, we may be drawn towards particular objects, like a flash of bioluminescence in the sea or the structural colour of a butterfly's wing, and away from others, like the brown-pigmented vole.

That the eye is not perfect is evidence *towards* the theory of evolution. As this book reveals through its various case studies, the imperfections or limitations become selection pressures to other animals, which have evolved to exploit them. Colour is the ideal subject to demonstrate this and consequently to help explain evolution, since it is purely a figment of our imagination – the part of the visual process that Darwin overlooked. The tree frog and milk snake appear green to the animals that matter, yet impart no wavelengths for green whatsoever.

Given the real facts, Darwin would have known a blue and yellow frog, and a pink snake. Instead, his assumption that these animals were simply green and conspicuously coloured respectively told him nothing of the problems associated with eyes in the environment, and sums up well his error. The colour factories within animal bodies, which emanate electromagnetic rays into the environment, are just as sophisticated as the eyes of ourselves and of other animals. Animal colours, it has been shown, are a matter of life or death. A visual arms race exists, and often an arms race *is* evolution, as Darwin sometimes portrayed it in *The Origin*. I hope this book can put the controversy of *perfection* to rest. Although how a *complex* organ such as an eye can *result* from the process of evolution, via 'graduations' (in Darwin's words), is an altogether different problem – one similar to how animals could *all* 'quickly' evolve hard parts when called upon in the Cambrian period. This is the subject for another book . . .

Suggested Reading

Introduction

Darwin, C., 1894, *The Origin of Species*, sixth edition, John Murray, London.
Land, M. F. & Nilsson, D. E., 2002, *Animal Eyes*, Oxford University Press, Oxford.
Newton, I., 1740, *Opticks*, reprinted from the fourth edition by Dover Publications Inc., New York.
Wright, W. D., 1963, 'The rays are not coloured', *Nature*, **198**, 1239–1244.

Ultraviolet

Cianci, M. & others, 2002, 'The molecular basis of the coloration mechanism in lobster shell: ß-Crustacyanin at 3.2-Å resolution', *Proceedings of the National Academy of Sciences*, **99**, 9795–9800.
Fox, H. M. & Vevers, G., 1960,*The Nature of Animal Colours*, Sidgwick & Jackson Ltd., London.
Fox, R., Lehmkuhle, S. W. & Westendorf, D. H., 1976, 'Falcon visual acuity', *Science*, **192**, 263–265.
Koivula, M. & Viitala, J., 1999, 'Rough-legged buzzards use vole scent marks to assess hunting areas', *Journal of Avian Biology*, **30**, 329–332.
Verne, J., 1930, *Couleurs et pigments des êtres vivants*, Armand Colin, Paris.
Viitala, J. & others, 1995, 'Attraction of kestrels to vole scent marks', *Nature*, **373**, 425–427.
Withgott, J., 2001, 'Feeling the burn', *Natural History*, July–August, 38–44.

Violet

Goureau, M., 1842, 'Sur l'irisation des ailes des insectes', *Annuaire de la Société Entomologique de France* **12**, 201–215.

Ghiradella, H., 1989, 'Structure and development of iridescent butterfly scales: lattices and laminae', *Journal of Morphology* 202, 69–88.

Ghiradella, H., Aneshansley, D., Eisner, T., Silverglied, R. E. & Hinton, H. E., 1972, 'Ultraviolet reflection of a male butterfly: interference colour caused by thin-layer elaboration of wing scales', *Science* 178, 1214–1217.

Hooke, R., 1665, *Micrographia*, Martyn & Allestry, London.

Hutley, M. C., 1982, *Diffraction gratings*, Academic Press, London.

Land, M. F., 1972, 'The physics and biology of animal reflectors', *Progress in Biophysics and Molecular Biology* 24, 75–106.

Mason, C. W., 1927, 'Structural colours in insects' II and III, *Journal of Physical Chemistry*, 31, 321–354, 1856–1872.

Parker, A. R., 2000, '515 million years of structural colour', *Journal of Optics A: Pure and Applied Optics* 2, R15–28.

Parker, A. R., McPhedran, R. C., McKenzie, D. R., Botten, L. C. and Nicorovici, N.-A. P., 2001, 'Aphrodite's iridescence', *Nature* 409, 36–37.

Parker, A. R., Welch, V. L., Driver, D & Martini, N., 2003, 'An opal analogue discovered in a weevil', *Nature* 426, 786–787.

Raman, C. V., 1934, 'The origin of the colours in the plumage of birds', *Proceedings of the Indian Academy of Sciences* A1, 1–7.

Blue

Harvey, E. N., 1952, *Bioluminescence*, Academic Press Inc., New York.

Hastings, J. W., 1976, 'Bioluminescence', *Oceanus*, Winter, 17–27.

Herring, P. J. 1987, 'Systematic distribution of bioluminescence in living organisms', *Journal of Bioluminescence and Chemiluminescence*, 1, 147–163.

Morin, J. G. & others, 1975, 'Light for all reasons: variations in the behavioural repertoire of the flashlight fish', *Science*, 190, 74–76.

Green

Fox, D. L., 1976, *Animal Biochromes and Structural Colours*, University of California Press, Berkeley.

McCarney, E. S., 1976, *Optics of the Atmosphere: Scattering by Molecules and Particles*, Wiley, New York.

Stachowicz, J. J. & Hay, M. E., 2000, 'Geographic variation in camouflage specialization by a decorator crab', The *American Naturalist*, 156, 59–71.

Yellow

Arnold, K. E. & others, 2002, 'Fluorescent signalling in parrots', *Science*, 296, 92 (contrast with the paper by Pearn and others).

Mitter, P., 2001, *Indian Art*, Oxford University Press, Oxford.

Nemésio, A., 2001, 'Colour production and evolution in parrots', *International Journal of Ornithology* 4, 75–102.

Ormo, M. & others, 1996, 'Crystal structure of the *Aequorea victoria* green fluorescent protein', *Science*, 273, 1392–1395.

Parker, A. R., 2002, 'Fluorescence of yellow budgerigars', *Science* 296, 655.

Pearn, S. M., Bennett, A. T. D. & Cuthill, I. C., 2001, 'Ultraviolet vision, fluorescence and mate choice in a parrot, the budgerigar Melopsittacus undulatus', Proceedings of the Royal Society of London, Biological Sciences, 268, 2273–2279.

Prum, R. O. & others, 1998, 'Coherent light scattering by blue feather barbs', Nature, 396, 28–29.

Verrell, P. A., 1991, 'Illegitimate exploitation of sexual signalling systems and the origin of species', Ethology, Ecology and Evolution 3, 273–283.

Orange

Benson, W. W., 'Natural selection for Müllurian mimicry in Heliconius erato in Costa Rica', Science, 176, 936–939.

Kapan, D. D. & McDiarmid, R. W., 1981, 'Coral snake mimicry: does it occur?' mimicry', Nature, 409, 338–340. 1212.

Kaufman, L. & Rock, I., 1962, 'The moon illusion', Scien..., ...te mimics', Nature, 418, 524–526.

Norman, M. D., Finn, J. & Tregenza, T., 2001, 'Dynamic mimicry in an ...field test of müllerian Malayan Octopus', Proceedings of the Royal Society of London, Biological Sciences, 268, 1755–1758. 1996, Cephalopod Behaviour, Cambridge ...ly, 1–10.

Parker, G. H., 1948, Animal Colour Changes and their Neurohumours, Cambridge University Press, Cambridge.

Wickler, W., 1968, Mimicry in Plants and Animals, Weidenfeld & Nicolson, London.

Red

Douglas, R. H., Mullineaux, C. W. & Partridge, J. C., 2000, 'Long-wave sensitivity in deep-sea stomiid dragonfish with far-red bioluminescence: evidence for a dietary origin of the chlorophyll-derived retinal photosensitizer of Malacosteus niger', Philosophical Transactions of the Royal Society of London: Biological Sciences, 355, 1269–1272.

Herring, P. J., 2002, The Biology of the Deep Ocean, Oxford University Press, Oxford.

Horváth, G., & Varjú, D., 2004, Polarized Light in Animal Vision, Springer-Verlag, Berlin.

Lohse, D., Schmitz, B. & Versluis, M., 2001, 'Snapping shrimp make flashing Bubbles', Nature, 413, 477–478.

Miller, W. H., Moller, A. R. & Bernhard, C. G., 1966, 'The corneal nipple array', The functional organisation of the compound eye (C. G. Bernhard, ed.), 21–33, Pergamon Press, Oxford.

Parker, A. R., 1999, 'Light-reflection strategies', American Scientist 87, 248–255.

Parker, A. R., Hegedus, Z. & Watts, R. A., 1998, 'Solar-absorber type anti-reflector on the eye of an Eocene fly (45Ma)', Proceedings of the Royal Society of London: Biological Sciences 265, 811–815.

Large, M. C. J. & others, 2001, 'The mechanism of light reflectance in silverfish', Proceedings of the Royal Society of London: Mathematical and Physical Sciences 457, 511–518.

Land, M. F., 1978, 'Animal eyes with mirror optics', *Scientific American* 239, 126–134.

Denton, E. J., 1970, 'On the organization of reflecting surfaces in some marine animals', *Philosophical Transactions of the Royal Society of London: Biological Sciences* 258, 285–313.

Denton, E. J., 1990, 'Light and vision at depths greater than 200 metres', *Light and Life In the Sea* (P. J. Herring, A. K. Campbell, M. Whitfield & L. Maddock, eds), Cambridge University Press, Cambridge, 127–148.

Conclusion for Evolution

Briggs, D. E. G., Erwin, D. H. & Collier, F. J., 1994, ... the effect of light on Smithsonian Institution Press, Washington, *the Royal Society of London*:

Gould, S. J., 1989, *Wonderful Life*, W....

Parker, A. R., 1998, 'Colour ..... evolution in the Ca Blink of an Eye*. The Free Press, London. *Biological Scie...

Parker, A. R. n. & others, 2004, *The Cambrian Fossils of Chengjiang, China*, Xian-G...well Publishing, Oxford.

Index

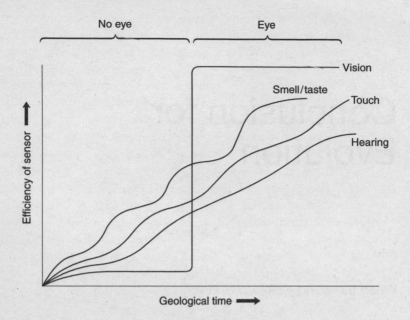

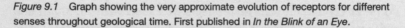

Figure 9.1 Graph showing the very approximate evolution of receptors for different senses throughout geological time. First published in *In the Blink of an Eye*.

Earth. I have reproduced a figure from that book above. It is the simplest, crudest diagram in the whole book, but one I consider the most important for the message it holds.

The figure demonstrates that vision is the only sense that can appear in the history of life within a very short period of geological time. Other senses, such as smell, touch and hearing, staggered into the history of life through gradually improving the efficiency of their sensors. They could not have unleashed a bombshell for animal life on Earth in general; the type of bombshell that could cause that life to explode into new body forms with hard parts. Vision, on the other hand, is a sense that exists either efficiently or not at all.

'Vision' demands an *image-forming* eye. This is not a minor detail, nor is it one simple to comprehend, and has been misinterpreted on numerous occasions. A light receptor that detects only light intensity,

and mate choice in a parrot, the budgerigar *Melopsittacus undulatus*', *Proceedings of the Royal Society of London, Biological Sciences*, 268, 2273–2279.

Prum, R. O. & others, 1998, 'Coherent light scattering by blue feather barbs', *Nature*, 396, 28–29.

Verrell, P. A., 1991, 'Illegitimate exploitation of sexual signalling systems and the origin of species', *Ethology, Ecology and Evolution* 3, 273–283.

Orange

Benson, W. W., 'Natural selection for Müllurian mimicry in *Heliconius erato* in Costa Rica', *Science*, 176, 936–939.

Greene, H. W. & McDiarmid, R. W., 1981, 'Coral snake mimicry: does it occur?' *Science*, 213, 1207–1212.

Hanlon, R. T. & Messenger, J. B., 1996, *Cephalopod Behaviour*, Cambridge University Press, Cambridge.

Johnstone, R. A., 2002, 'The evolution of inaccurate mimics', *Nature*, 418, 524–526.

Kapan, D. D., 2001, 'Three-butterfly system provides a field test of müllerian mimicry', *Nature*, 409, 338–340.

Kaufman, L. & Rock, I., 1962, 'The moon illusion', *Scientific American*, July, 1–10.

Norman, M. D., Finn, J. & Tregenza, T., 2001, 'Dynamic mimicry in an Indo-Malayan Octopus', *Proceedings of the Royal Society of London, Biological Sciences*, 268, 1755–1758.

Parker, G. H., 1948, *Animal Colour Changes and their Neurohumours*, Cambridge University Press, Cambridge.

Wickler, W., 1968, *Mimicry in Plants and Animals*, Weidenfield & Nicolson, London.

Red

Douglas, R. H., Mullineaux, C. W. & Partridge, J. C., 2000, 'Long-wave sensitivity in deep-sea stomiid dragonfish with far-red bioluminescence: evidence for a dietary origin of the chlorophyll-derived retinal photosensitizer of *Malacosteus niger*', *Philosophical Transactions of the Royal Society of London: Biological Sciences*, 355, 1269–1272.

Herring, P. J., 2002, *The Biology of the Deep Ocean*, Oxford University Press, Oxford.

Horváth, G., & Varjú, D., 2004, *Polarized Light in Animal Vision*, Springer-Verlag, Berlin.

Lohse, D., Schmitz, B. & Versluis, M., 2001, 'Snapping shrimp make flashing Bubbles', *Nature*, 413, 477–478.

Miller, W. H., Moller, A. R. & Bernhard, C. G., 1966, 'The corneal nipple array', *The functional organisation of the compound eye* (C. G. Bernhard, ed.), 21–33, Pergamon Press, Oxford.

Parker, A. R., 1999, 'Light-reflection strategies', *American Scientist* 87, 248–255.

Parker, A. R., Hegedus, Z. & Watts, R. A., 1998, 'Solar-absorber type anti-reflector on the eye of an Eocene fly (45Mu)', *Proceedings of the Royal Society of London: Biological Sciences* 265, 811–815.

Large, M. C. J. & others, 2001, 'The mechanism of light reflectance in silverfish', *Proceedings of the Royal Society of London: Mathematical and Physical Sciences* 457, 511–518.

Land, M. F., 1978, 'Animal eyes with mirror optics', *Scientific American* **239**, 126–134.

Denton, E. J., 1970, 'On the organization of reflecting surfaces in some marine animals', *Philosophical Transactions of the Royal Society of London: Biological Sciences* **258**, 285–313.

Denton, E. J., 1990, 'Light and vision at depths greater than 200 metres', *Light and Life In the Sea* (P. J. Herring, A. K. Campbell, M. Whitfield & L. Maddock, eds.), Cambridge University Press, Cambridge, 127–148.

Conclusion for Evolution

Briggs, D. E. G., Erwin, D. H. & Collier, F. J., 1994, *The Fossils of the Burgess Shale*, Smithsonian Institution Press, Washington, D.C.

Gould, S. J., 1989, *Wonderful Life*, W.W. Norton & Co., New York.

Parker, A. R., 1998, 'Colour in Burgess Shale animals and the effect of light on evolution in the Cambrian', *Proceedings of the Royal Society of London: Biological Sciences*, **265**, 967–972.

Parker, A. R., 2003, *In the Blink of an Eye*. The Free Press, London.

Xian-Guang, H. & others, 2004, *The Cambrian Fossils of Chengjiang, China*, Blackwell Publishing, Oxford.

Index

Page numbers in *italic* refer to illustrations and figures in the text